Helen McGuinness

Anatomy & Physiology
WORKBOOK AND REVISION GUIDE

HODDER
EDUCATION
LEARN MORE

Contents

How to use this workbook

This anatomy and physiology workbook and revision guide is designed to be used as a companion to the *Anatomy & Physiology* textbook written by Helen McGuinness.

The workbook is filled with activities that are designed to test your knowledge of anatomy and physiology in a range of different ways, including:

- **multiple-choice questions** to test your understanding quickly and efficiently
- **exam-style short answer questions** designed to test accurate and concise key written knowledge
- **labelling activities of key diagrams** to test knowledge of structures and their location in the body
- **sequencing activities**, such as sorting into the correct order or category, to test knowledge of anatomy and physiology processes.
- **key word matching** to test knowledge of key words and terms
- **fill in the blanks** to test your knowledge of key subjects.

The activities can be attempted in any order and are designed to be completed to suit your particular style of study. Using a range of activities will help you to assess your knowledge in a comprehensive way when preparing for exams and assessments, and can provide valuable evidence of your understanding.

Before attempting the workbook activities, we recommend that you:

- review each chapter and its **revision summary**
- review the **key words** found towards the end of the chapter.

The answers and some extra activities can be found online at www.hoddereducation.co.uk/Anatomy-and-Physiology-Extras

Chapter 1 An introduction to anatomy and physiology: how the body is organised

INTRODUCTION

The human body may be likened to a map – in order to locate parts of the body correctly, students should be familiar with directional terms and references.

The essentials you need for exams and assessments

You need to know:

- **anatomical directional terminology**, which give a precise description of a body part, for example 'anterior' and 'posterior'
- **anatomical positions** or planes, which divide the body into sections, for example 'frontal', 'median' and 'transverse'
- **anatomical regional terms**, which refer to specific areas of the body, for example 'axillary' and 'cephalic'.

Note that anatomical terms relating to movement are covered in Chapter 4, the skeletal system, on page 17.

ACTIVITY 1: MULTIPLE-CHOICE QUESTIONS

1 The anatomical term 'peripheral' means:
   a at or near the centre
   b away from the centre
   c away from the head
   d relating to the head end.

2 The anatomical term relating to the palm-side of the hand is:
   a plantar
   b palmar
   c prone
   d proximal.

3 'Inguinal' is an anatomical term relating to which area of the body?
   a buttocks
   b lower back
   c groin
   d navel

4 The cephalic region of the body is located in the:
   a head
   b neck
   c chest
   d ribs.

5 The anatomical term describing the area of the body relating to the leg/thigh is:
   a calcaneal
   b carpal
   c crural
   d sural.

6 When describing a body part that is on the opposite side to another structure, the correct term to use is:
   a caudal
   b ipsilateral
   c visceral
   d contralateral.

7 'Coeliac' is an anatomical regional term for the:
   a upper limbs
   b head and neck
   c abdomen
   d lower limbs.

8 'Axillary' is a term relating to the:
   a abdomen
   b armpit
   c breast
   d ribs.

9 The teeth are located in the:
   a otic cavity
   b ophthalmic cavity
   c oral cavity
   d orbital cavity.

10 'Parietal' is a term is used to refer to things within the body that are attached to the:
   a length of a body part
   b internal organs
   c outside of a body cavity
   d inner walls of a body cavity.

Activity 2: Exam-style questions

1 Describe the standardised body position referred to as the anatomical position.   **4 marks**
2 State the three anatomical planes that separate the body into sections.   **3 marks**
3 a Define the following anatomical terms.   **8 marks**
   i Superior   ii Supine   iii Lateral   iv Proximal
  b Define the following anatomical regional terms, stating which region they are in.   **8 marks**
   i Buccal   ii Inguinal   iii Cubital   iv Femoral
4 State the two main body cavities and describe where they are located.   **4 marks**

ACTIVITY 3: LABELLING THE ANATOMICAL REGIONS

Label the anatomical regions using the words shown below. Note that some labels need to be used twice.

Abdominal	Axillary	Brachial	Calcaneal	Carpal	Costal	Crural
Cubital	Digital/phalangeal	Femoral	Forearm	Gluteal	Inguinal	Lumbar
Mammary	Palmar	Patellar	Pectoral	Pedal	Pelvic	Pericardial
Perineal	Plantar	Popliteal	Sacral	Sural		Thoracic
Umbilical	Vertebral					

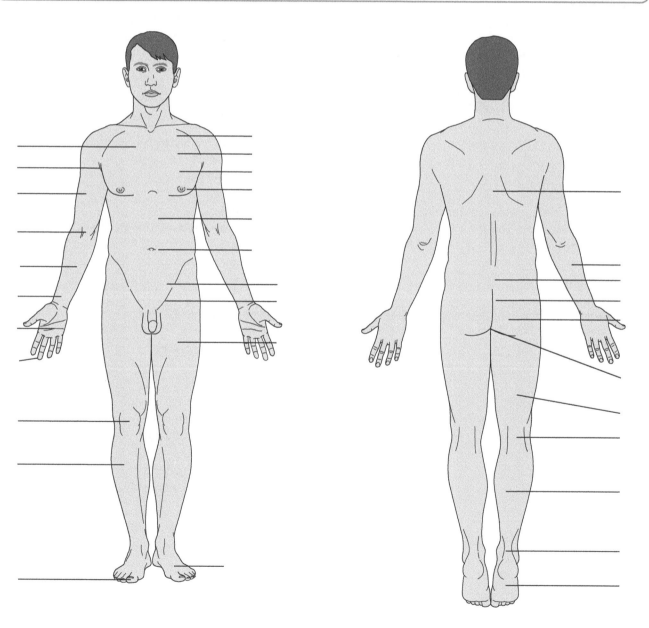

ACTIVITY 4: MATCH THE KEY WORDS

Anatomical directional terminology is used to give a precise description of a body part.
Match the following key terms to the correct description.

Deep   Visceral   Caudal   Proximal   Ipsilateral
Superior   Lateral   Prone   Anterior   Plantar

Key term	Description	Key term	Description
	Away from the mid-line		Away from the head, or below
	Further from the body surface		Lying face down in a horizontal position
	Front surface of the body/ structure		Situated towards the head, or above
	Nearest to the point of reference		Term used when referring to any internal organs
	On the same side as another structure		Relating to the sole of the foot

ACTIVITY 5: FILL IN THE BLANKS

Test your knowledge of anatomical regional terms by filling in the gaps with the terms shown below.

axillary   breasts   cheek   femoral   forearm
forehead   lumbar   neck   ribs   wrist

1   The vertebrae in the lower back are the _____.
2   The _____ lymph nodes are situated under the arm.
3   The radius and ulna are bones of the _____.
4   The frontal bone is found in the _____.
5   The mammary glands are also known as the _____.
6   The carpal bones are located in the _____.
7   The buccal cavity is in the _____.
8   The costal cartilages are between the _____.
9   The _____ artery is in the thigh.
10 The bones of the _____ are called the cervical vertebrae.

Chapter 2 Cells and tissues

INTRODUCTION

It is important for therapists to have an understanding of cells and tissues as they are the building blocks upon which the human body is formed.

Knowing how the body functions at a cellular level helps therapists to understand how the body functions in both health and illness, as well as to understand the link between structure and function.

An understanding of tissues is essential to bridge your knowledge between cells and the organisation of the body's organs.

The essentials you need for exams and assessments

You need to know:

- **levels of structural organisation in the body**: chemical, cellular, tissues, organs and systems
- **structure of a cell**: be able to label and understand the parts of a cell, for example cell membrane and nucleus
- **characteristics/functions of a cell**: know the variety of functions that cells carry out in order to survive (for example respiration and excretion) and the cellular functions used to activate the energy needed for the cell to function (cellular transport, such as osmosis and diffusion) and cellular metabolism (anabolism and catabolism)
- **cell division**: how cells divide and reproduce in order to maintain life (mitosis and meiosis)
- **main types of tissues in the body**: epithelial, connective, nervous, muscular and membranes, along with examples of where these may be found in the body.

ACTIVITY 1: MULTIPLE-CHOICE QUESTIONS

1 The process by which new body cells are produced for growth and repair is called:
  a meiosis
  b metaphase
  c mitosis
  d metabolism.

2 The nucleus of a human cell contains:
  a 43 chromosomes
  b 46 chromosomes
  c 26 chromosomes
  d 23 chromosomes.

3 The final stage of mitosis is known as:
  a anaphase
  b telophase
  c prophase
  d metaphase.

4 The control centre of the cell that directs nearly all metabolic activities is the:
  a mitochondrion
  b Golgi body
  c nucleus
  d cell membrane.

5 The organelle that powers cell activities is the:
  a nucleus
  b lysosome
  c mitochondrion
  d endoplasmic reticulum.

6 The function of a ribosome in a cell is to:
  a destroy bacteria
  b manufacture protein
  c supply energy
  d dispose of waste.

7 The process by which small molecules move from areas of high concentration to those of lower concentration is:
  a osmosis
  b diffusion
  c filtration
  d active transport.

8 Catabolism is the:
  a basic chemical working of the body cells
  b building up of complex molecules
  c rate at which a person consumes energy in activities
  d chemical breakdown of complex substances by the body to form simpler ones.

9 The type of tissue that lines the internal and external organs of the body, as well as vessels and body cavities, is:
   a connective tissue
   b serous tissue
   c epithelial tissue
   d nervous tissue.

10 Membranes that line openings to the outside of the body are called:
   a serous membranes
   b mucous membranes
   c synovial membranes
   d epithelial membranes.

Activity 2: Exam-style questions

1 State the function of the following cell organelles.   **2 marks**
   a Mitochondria
   b Ribosomes
2 Describe the two ways in which cells divide.   **2 marks**
   a Mitosis
   b Meiosis
3 Describe the following methods of cell transport.   **2 marks**
   a Osmosis
   b Diffusion
4 Name three places where mucous membranes are located in the body.   **3 marks**
5 Name two places where keratinised stratified epithelium is located in the body.   **2 marks**

ACTIVITY 3: LABELLING THE STRUCTURE OF A CELL

Label the structure of a cell using the words shown below.

Cell membrane	Centriole	Chromatin	Cytoplasm
Golgi body	Lysosome	Mitochondrion	Nuclear membrane
Nucleolus	Nucleus	Ribosome	Rough endoplasmic reticulum
Smooth endoplasmic reticulum	Vacuole		

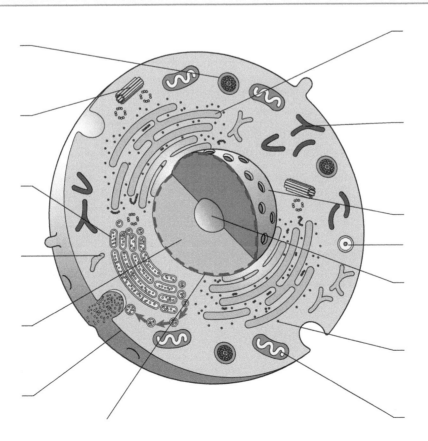

ACTIVITY 4: SORTING

The human body has five levels of structural organisation. Put the following into the correct order, starting from the lowest level of organisation.

System level   Cellular level   Chemical level (atoms and molecules)   Organ level   Tissue level

1	
2	
3	
4	
5	

ACTIVITY 5: MATCH THE KEY WORDS

You need to understand the functional significance of the cell's structural parts.
Match the following key terms to the correct description.

Cytoplasm   Centrioles   Golgi body   Chromosomes   Endoplasmic reticulum
Mitochondria   Cell membrane   Vacuole   Ribosomes   Nucleus

Key term	Description	Key term	Description
	Oval-shaped organelles that provide the energy to power the cell's activities		A series of membranes continuous with the cell membrane
	Empty spaces within the cytoplasm that contain waste materials		A collection of flattened sacs within the cytoplasm
	Small spherical structures that are associated with cell division		A gel-like substance that is enclosed by the cell membrane
	Tiny organelles that help to manufacture protein for use within the cell		A fine film that encloses the cell and protects its contents
	Control centre of the cell; regulates the cell's functions		Thread-like structures in the cell nucleus that carry genetic information

ACTIVITY 6: FILL IN THE BLANKS

Test your knowledge of cellular respiration by filling in the gaps with the terms shown below.

active	diffusion	energy	equilibrium	extracellular	filtration	glucose
homeostasis	nutrients	osmosis	phagocytosis	pinocytosis	waste	

A cell must maintain a stable internal environment in order to function properly. The process by which the body maintains a stable internal cellular environment is called _____.

To sustain life, many substances need to be transported into, out of, and between cells. The term 'cell respiration' refers to the controlled exchange of _____ (such as oxygen and glucose) and _____ (such as carbon dioxide) by the cell to activate the energy needed for the cell to function. Substances are absorbed, or excreted, through the cell membrane in several different ways. _____ is the process by which small molecules move from an area of high concentration to lower concentration. It is the basis by which the cells lining the small intestines take in digestive products to be used by the body.

_____ refers to the movement of water through the cell membrane from areas of low chemical concentration to areas of high chemical concentration. This allows for the dilution of chemicals, which are unable to cross the cell membrane by diffusion, in order to maintain _____ within the cell. The movement of water and dissolved substances across the cell membrane due to differences in pressure is called _____; an example of this is in the kidneys.

_____ transport is used when cells needs to transport substances against their concentration gradient. This is an _____ dependent process in which the cell takes in larger molecules that would be otherwise unable to enter in sufficient quantities. This process is the means by which the cell absorbs _____.

The method by which a cell absorbs small particles outside the cell and brings them inside is called _____; this process is usually used for taking in _____ fluid.

_____ is the process by which a cell engulfs particles. such as bacteria, other micro-organisms and foreign matter.

Chapter 3 The skin, hair and nails

INTRODUCTION

Skin is a dynamic organ: it is in a constant state of change as the cells of the outer layers are continuously shed and replaced by inner cells moving up to the surface. It acts like a cell membrane, providing a barrier between the external environment and our internal organs.

It is important for therapists to have a comprehensive knowledge of the structure and functions of skin in order to understand the processes of cell renewal and product penetration, as well as being able to offer the most effective treatments and products for their client's skin type.

The skin is significant in that it reflects the health and wellbeing of clients, and it is the foundation upon which all treatments are undertaken.

The essentials you need for exams and assessments

You need to know:

- **Structure of the skin:**
  - **Epidermis**: outer, most superficial layer.
    - Five layers: stratum corneum, the outermost (horny layer); stratum lucidum (transparent layer); stratum granulosum (granular layer); stratum spinosum (prickle cell layer); and stratum germinativum, the innermost (basal layer).
    - Cells present: melanocytes, keratinocytes, Langerhan cells.
  - **Dermis**: inner, deeper layer that provides structural support.
    - Consists of papillary and reticular layers containing fibroblasts that produce collagen, elastin and glycosaminoglycan; other cells present in dermis include mast cells (histamine) and macrophages, along with the blood supply and lymphatic vessels.
    - Contains appendages such as hair, hair follicles, sebaceous glands, arrector pili muscles, dermal papillae, sweat glands (eccrine and apocrine), sweat pores, sweat ducts, and sensory and motor nerves.
  - **Subcutaneous layer**: insulating, protective layer below dermis; made up of adipose tissue.
- **Growth stages of the skin:**
  - **Stage 1**: cell regeneration/mitosis in stratum germinativum of epidermis.
  - **Stage 2**: cell maturation – new cells mature as they migrate from the stratum germinativum to the stratum spinosum.
  - **Stage 3**: keratinisation – the cells undergo change and die in the stratum granulosum.
  - **Stage 4**: desquamation – the cells reach the end of their life and are sloughed off.
- **Repair stages of the skin:**
  - **Stage 1**: haemostasis
  - **Stage 2**: inflammation
  - **Stage 3**: proliferation
  - **Stage 4**: maturation.
- **Functions of the skin (SHAPES VM)**: secretion, heat regulation, absorption, protection, excretion, sensation, vitamin D formation, melanin formation.
- **Skin types**: normal, dry, oily, combination.
- **Skin conditions**: sensitive, dehydrated, mature.
- **Structure of the hair**: hair shaft (layers from outside in are cuticle, cortex and medulla); inner root sheath (cuticle, Huxley and Henle layer), outer root sheath, connective sheath, vitreous membrane, hair bulb, matrix.
- **Hair growth cycle**: anagen (active growing stage), catagen (transitional stage) and telogen (resting stage).
- **Hair types of the body**: including lanugo, vellus and terminal.
- **Factors that affect hair growth**: including congenital, hormonal, topical, climate, stress, medication and illness.
- **Structure of the nail**: including nail plate, nail bed, nail grooves, matrix, cuticle (eponychium, perionychium, hyponychium), nail mantle/proximal nail fold, lunula, nail wall and free edge.

- **Process of nail growth:**
  - Nail growth occurs from the nail matrix by cell division.
  - As new cells are produced in the matrix, older cells are pushed forward and are hardened by the process of keratinisation, which forms the hardened nail plate.
  - It takes approximately six months for cells to travel from the lunula to the free edge of the nail for fingernails, and approximately nine to twelve months for toenails.
  - Growth is approximately 3 mm per month for fingernails (toe nail growth is slower).
- **Factors that affect nail growth:** including health, lifestyle, diet, age, climate, illness, medication, nail damage, skin texture, chemotherapy, radiotherapy, smoking, alcohol, stress, lack of sleep, current hand and nail care routine.

Skin

ACTIVITY 1: MULTIPLE-CHOICE QUESTIONS

Skin

1 In which of the following layers are epidermal cells constantly being reproduced?
   a stratum corneum
   b stratum granulosum
   c stratum germinativum
   d stratum lucidum

2 The thickest layer of the epidermis is the:
   a stratum spinosum
   b stratum granulosum
   c stratum corneum
   d stratum germinativum.

3 The function of the sebaceous gland is to:
   a secrete sweat
   b secrete sebum
   c regulate temperature
   d insulate.

4 Which of the following is responsible for protecting the deeper layers of the skin from UV damage?
   a keratin
   b sebum
   c carotene
   d melanin

5 A Merkel's disk is a type of cutaneous receptor used to detect:
   a deep pressure, fast vibrations
   b changes in texture, slow vibrations
   c sustained touch and pressure
   d pain arising from mechanical stimuli.

6 The term used to describe when living skin cells change to dead cells is:
   a keratinisation
   b desquamation
   c mitosis
   d maturation.

7 Sebum contains a mixture of:
   a collagen and elastin
   b sodium and electrolytes
   c fats, cholesterol and cell debris
   d fats and water.

8 The cells responsible for laying down and maintaining the extracellular matrix and the structure of the dermis are:
   a mast cells
   b fibroblasts
   c Langerhan cells
   d phagocytes.

9 In which layer of the dermis would you find collagen fibres?
   a papillary layer
   b basal cell layer
   c subcutaneous layer
   d reticular layer

10 The function of the subcutaneous layer is to:
   a support blood vessels
   b insulate the body
   c support nerve endings
   d all of the above.

11 The type of sweat glands that are widely distributed throughout the body are:
   a apocrine
   b eccrine
   c adipose
   d sebaceous.

12 A function of keratin in the skin is to provide:
   a waste elimination
   b a blood supply
   c protection
   d nourishment.

Skin type and skin conditions

1   The main recognition factors of an oily skin include:
    a   thick, coarse and congested with enlarged pores
    b   thin, coarse and flaky with enlarged pores
    c   thick, coarse and congested with fine pores
    d   thick, smooth and congested with enlarged pores.

2   A dry skin is so-called because it is lacking in:
    a   elasticity
    b   colour
    c   thickness
    d   sebum.

3   What is meant by the term 'erythema' in the skin?
    a   a mass of dilated capillaries in the skin
    b   a small, raised elevation in the skin
    c   a mark left on the skin after a wound has healed
    d   reddening of the skin due to the dilation of blood capillaries

4   Sensitivity is a skin condition that may be due to:
    a   a reaction to specific product ingredients
    b   misuse of products
    c   diet
    d   all of the above.

5   A small raised elevation of the skin is known as a:
    a   pustule
    b   papule
    c   nodule
    d   macule.

6   The skin type likely to present with the best elasticity is:
    a   dry
    b   oily
    c   sensitive
    d   mature.

7   Which of the following skin tones is most likely to be affected by the condition dermatosis papulose nigra (DPN)?
    a   brown
    b   black
    c   white
    d   olive

8   A milia is:
    a   a pearly, hard white nodule under the skin
    b   an abnormal sac containing fluid
    c   a crack in the epidermis
    d   a small, flat patch of increased pigmentation.

Skin diseases and disorders

1   Chloasma is best described as:
    a   an inflammatory skin condition
    b   a malignant tumour
    c   a pigmentation disorder
    d   a benign growth on the skin.

2   The most lethal type of skin cancer is:
    a   squamous cell carcinoma
    b   rodent ulcer
    c   basal cell carcinoma
    d   malignant melanoma.

3   A chronic inflammatory skin disease in which the skin appears abnormally red is known as:
    a   seborrhoea
    b   acne vulgaris
    c   rosacea
    d   folliculitis.

4   A sebaceous cyst appears as:
    a   a round, nodular lesion with a smooth, shiny surface
    b   an excessive secretion of sebum
    c   an excessive production of sweat
    d   a small inflamed nodule around a hair follicle.

5   The genetic chronic inflammatory skin disease that causes skin cells to reproduce too quickly is called:
    a   hyperkeratosis
    b   psoriasis
    c   lupus
    d   seborrhoeic dermatitis.

6   The main characteristic of a malignant melanoma is:
    a   a hard and warty appearance
    b   a blue-black nodule that increases in size, shape and colour
    c   a slow-growing pearly nodule growing at the site of a previous injury
    d   a collection of dilated capillaries radiating from a central papule.

7   Which of the following is a fungal infection in the skin?
    a   ringworm
    b   impetigo
    c   herpes simplex
    d   herpes zoster

ACTIVITY 2: LABELLING THE STRUCTURE OF THE SKIN

Label the cross-sectional diagram of the skin using the words shown below.

Artery	Capillary network	Cold receptor (Krause corpuscle)
Deep fascia	Dermis	Epidermis
Erector pili muscle	Hair	Hair bulb
Hair follicle	Heat receptor (Ruffini endings)	Motor nerve
Nerve endings	Pacinian corpuscle (pressure receptor)	Pain receptor
Sebaceous gland	Subcutaneous fat	Subcutaneous layer
Subdermal muscle layer	Sweat (eccrine) gland	Touch receptor (Meissner's corpuscle)
Vein		

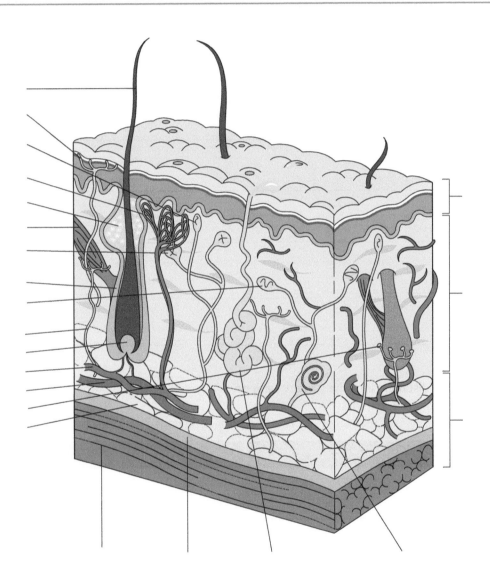

ACTIVITY 3: SORTING

The layers of the skin

Put the layers of the skin in the correct order, from the outermost layer down.

Papillary layer of dermis   Reticular layer of dermis   Stratum corneum   Stratum germinativum
Stratum granulosum   Stratum lucidum   Stratum spinosum   Subcutaneous layer

1	
2	
3	
4	
5	
6	
7	
8	

Skin disorders

Sort the following skin disorders into the correct category.

Chloasma   Conjunctivitis   Dermatitis   Eczema   Herpes simplex   Impetigo
Rosacea   Scabies   Seborrhea   Tinea corporis (ringworm)   Tinea pedis   Urticaria

Infectious	Non-infectious

ACTIVITY 4: MATCH THE KEY WORDS

Match the following key terms to the correct description.

Apocrine   Collagen   Corneocyte   Dermis   Epidermis   Extracellular matrix
Glycosaminoglycans   Keratin   Keratinisation   Papillary   Sebum   Stem

Key term	Description	Key term	Description
	Most superficial, thinner layer of skin		A type of cell found in the stratum germinativum of the epidermis; it is involved in the process of skin renewal
	A protein in the dermis responsible for elasticity and resilience		The uppermost layer of the dermis
	The process skin cells go through in which they change from living to dead		Key structural protein material making up hair, nails and the outer layer of skin
	An oily substance that coats the surface of skin and hair shafts		The support system of the dermis, made up of collagen, elastin and glycosaminoglycans
	The deeper, thicker layer of the skin		A dead skin cell of the stratum corneum
	Water-binding molecules found in the dermis that give the skin its plumpness		A type of sweat gland found in the genital and underarm regions

ACTIVITY 5: FILL IN THE BLANKS

Test your knowledge of skin types by filling in the gaps with the terms shown below.

blemishes   coarse   comedones   dry   elements   flaky   irritated   oil   open
pink   pores   pustules   reddens   sebum   shine   smooth   thin   water

Normal skin is balanced in that it has a good _____ and _____ balance. It has a _____ texture with a clear, even surface free from _____. _____ skin often has flaky patches, with small and tight pores due to lack of _____ production.

Oily skin is thick and _____, with large _____ pores and a characteristic _____. Blemishes are often very apparent in oily skins, along with blocked _____, _____, papules and _____ being present to varying degrees.

Sensitive skin presents with a _____ tone and appears _____ and translucent. It is prone to dry, _____ patches and is easily _____ by products and other external factors such as the _____. It _____ easily from any form of stimulation.

Hair

ACTIVITY 6: MULTIPLE-CHOICE QUESTIONS

1 Hair growth occurs from the:
   a cuticle
   b cortex
   c medulla
   d matrix.
2 Where are terminal hairs found in the body?
   a on the scalp
   b under the arms
   c in the pubic region
   d all of the above
3 Which of the following provides a crucial source of nourishment for hair?
   a dermal papilla
   b connective tissue sheath
   c outer root sheath
   d inner root sheath

4 Which of the following statements is **true** in relation to the hair growth cycle?
   a Catagen is a short resting stage.
   b Anagen lasts approximately two to four weeks.
   c Anagen is the active growing stage.
   d In anagen the hair separates from the dermal papilla and moves slowly up the follicle.
5 Hair colour is due to the presence of melanin in which layers of hair?
   a cortex and medulla
   b cuticle, cortex and medulla
   c cuticle and medulla
   d cuticle and cortex
6 The part of the hair structure that supplies the follicle with nerves and blood is the:
   a inner root sheath
   b outer root sheath
   c matrix
   d connective tissue sheath.

ACTIVITY 7: LABELLING THE STRUCTURE OF A HAIR

Label the cross-sectional diagram of a hair using the words shown below.

Connective tissue sheath	Dermal papilla	Erector pili muscle	Hair shaft
Inner root sheath	Matrix	Outer root sheath	Upper bulb

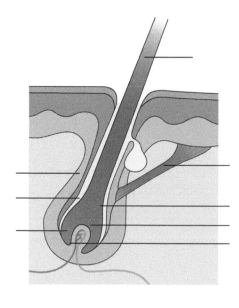

Nails

ACTIVITY 8: MULTIPLE-CHOICE QUESTIONS

1 The substance the nail is mainly composed of is:
   a calcium
   b keratin
   c blood
   d collagen.

2 The area of the nail where living cells are produced is the:
   a nail bed
   b lunula
   c matrix
   d nail wall.

3 The part of the nail that protects the matrix and provides a protective seal against bacteria is the:
   a cuticle
   b nail wall
   c nail groove
   d lunula.

4 The part of the nail that protects the matrix/nail root from physical damage is the:
   a nail plate
   b proximal nail fold
   c nail wall
   d nail bed.

5 The cuticle skin found under the free edge of the nail is known as the:
   a eponychium
   b perionychium
   c lateral nail fold
   d hyponychium.

6 A common nail disease characterised by inflammation and bacterial infection of the skin surrounding the nail is:
   a pterygium
   b onychomycosis
   c paronychia
   d leukonychia.

7 All nutrients are supplied to the nail via which layer of the skin?
   a the dermis
   b the epidermis
   c the subcutaneous layer
   d none of the above

8 The thickest layer of the nail plate, making up 70–75% of it, is:
   a the dorsal layer
   b the ventral layer
   c the intermediate layer
   d none of the above.

ACTIVITY 9: LABELLING A CROSS-SECTION OF A NAIL

Label the cross-sectional diagram of the nail using the words shown below.

Cuticle (eponychium)   Free edge   Hyponychium   Lunula   Nail bed
Nail grooves   Nail matrix   Nail plate   Nail wall   Proximal nail fold (nail mantle)

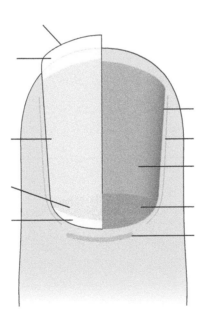

Mixed questions

Activity 10: Exam-style questions

1	In which layer of the skin are stem cells found?	**1 mark**
2	Name the two layers of the dermis.	**2 marks**
3	State two functions of the dermis.	**2 marks**
4	Name the cell responsible for making collagen and elastin fibres in the dermis.	**1 mark**
5	Name the type of tissue that makes up the subcutaneous layer of the skin.	**1 mark**
6	State the four sequential stages in skin repair.	**4 marks**
7	Describe two ways in which skin helps with temperature regulation.	**2 marks**
8	Describe two ways in which the skin acts as a protective organ.	**2 marks**
9	State two recognition factors of dry skin.	**2 marks**
10	State two key indicators of dehydrated skin.	**2 marks**
11	State two functions of hair in the body.	**2 marks**
12	State the three layers of the hair from the outer to inner layer.	**3 marks**
13 a	List the stages of hair growth in order.	**3 marks**
b	State the significance of each stage.	**3 marks**
14	What is the functional significance of the nail matrix?	**2 marks**
15 a	What is the functional significance of the cuticle in the nail?	**1 mark**
b	Describe the three areas of the cuticle.	**3 marks**
16	Name the hardest layer of the nail plate.	**1 mark**
17	Which layers of the epidermis form the nail plate?	**2 marks**
18	Which layers of the epidermis form the nail bed?	**3 marks**
19	State two recognition factors of onychomycosis.	**2 marks**
20	Describe the process of nail growth.	**3 marks**

Chapter 4 The skeletal system

INTRODUCTION

It is important for therapists to have a good working knowledge of the skeletal system. Bones are like landmarks in the body: by tracing their outlines you can be accurate in describing the position of muscles, glands and organs in the body.

The essentials you need for exams and assessments

You need to know:

- **position of bones of the axial skeleton**: skull, vertebral column, sternum and ribs
- **position of bones of the appendicular skeleton**: shoulder girdle, pelvic girdle and bones of the upper and lower limbs
- **structure and growth of a long bone**:
  - **structure**: compact, cancellous, parts of a long bone (including diaphysis, epiphysis, medullary cavity, epiphyseal cartilage and periosteum)
  - **bone growth**: ossification
- **types of bones** including long, short, irregular, flat and sesamoid, with examples
- **types of joints** in the body including fibrous (immovable), slightly movable or cartilaginous, freely movable or synovial, with examples
- **types of synovial joints** including pivot, hinge, condyloid, ball and socket, saddle and gliding, with examples
- **anatomical terms relating to movement** including flexion, extension, adduction, abduction, pronation, supination, dorsi flexion, plantar flexion, inversion and eversion
- **functions of the skeletal system** (PAM'S Skeleton Forces Movement).

ACTIVITY 1: MULTIPLE-CHOICE QUESTIONS

1 The type of joint that permits free movement is:
   a synovial
   b fibrous
   c cartilaginous
   d none of the above.

2 The long bone of the upper arm is the:
   a radius   c humerus
   b ulna   d occipital.

3 The bone positioned in the centre of the chest is the:
   a scapula   c clavicle
   b sternum   d sphenoid.

4 The vertebrae of the lower back are described as:
   a lumbar   c cervical
   b coccygeal   d sacral.

5 Phalanges are found in the:
   a toes only   c fingers and toes
   b fingers only   d ankle and wrist.

6 Which of the following statements is **true** in relation to a ligament?
   a A ligament joins bone to muscle.
   b A ligament is made up of articular cartilage.
   c Ligaments enable bones to move when muscles contract.
   d A ligament links bones together at joints.

7 A hinge joint permits:
   a rotation
   b adduction and abduction
   c flexion and extension
   d flexion, extension, rotation and circumduction.

8 An example of the location of a gliding joint is:
   a in the wrist
   b in the elbow
   c in the ankle
   d between the vertebrae.

9 The L-shaped bones that form the anterior part of the roof of the mouth are:
   a ethmoid
   b lacrimal
   c palatine
   d turbinate.

10 A joint disease characterised by the breakdown of articular cartilage is:
   a osteoarthritis
   b rheumatoid arthritis
   c bursitis
   d osteoporosis.

Activity 2: Exam-style questions

1 In the structure of a long bone, name the site of bone elongation during
   the growing years.   **1 mark**
2 Name the bones of the forearm.   **2 marks**
3 Name the bones of the shoulder girdle.   **2 marks**
4 Name the bones of the pelvic girdle.   **3 marks**
5 Give two examples of synovial joints and where they may be located.   **4 marks**
6 a Describe the following postural conditions.   **2 marks**
      i   Kyphosis
      ii Lordosis
   b In each case, describe the effects the postural conditions have on the body.   **2 marks**

ACTIVITY 3A: LABELLING THE BONES OF THE SKULL

Label the bones of the skull using the words shown below.

Frontal   Occipital   Parietal   Sphenoid   Temporal

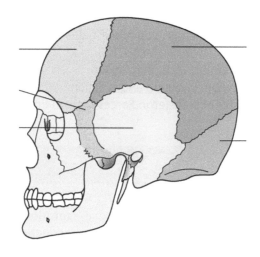

ACTIVITY 3B: LABELLING THE BONES OF THE FACE

Label the bones of the face using the words shown below.

Ethmoid   Lacrimal   Mandible   Maxilla
Nasal bone   Turbinate   Vomer   Zygomatic

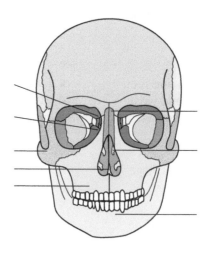

ACTIVITY 3C: LABELLING THE BONES OF CHEST, NECK AND SHOULDER

Label the bones of chest, neck and shoulder using the words shown below.

Cervical vertebrae   Clavicle   Humerus   Ribs   Scapula   Sternum

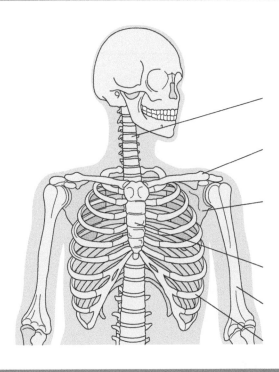

ACTIVITY 3D: LABELLING THE BONES OF THE ARM, HAND, LEG AND FOOT

Label the bones of the arm and hand (below left) using the words shown below.

Carpals   Humerus   Metacarpals   Phalanges   Radius   Ulna

Label the bones of the leg and foot (below right) using the words shown below.

Femur   Fibula   Metatarsals   Patella   Phalanges   Tarsals   Tibia

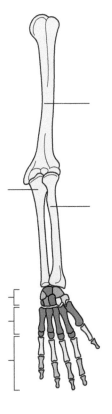

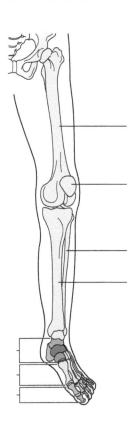

ACTIVITY 3E: LABELLING THE BONES OF THE SKELETON

Label the bones of the skeleton using the words shown below. Note that some labels need to be used twice.

Carpals	Cervical vertebrae	Clavicle	Coccyx	Femur	Fibula	Humerus	Ilium
Ischium	Lumbar vertebrae	Metacarpals	Metatarsals	Patella	Phalanges	Pubis	Radius
Ribs	Sacrum	Scapula	Skull	Sternum	Tarsals	Thoracic vertebrae	
Tibia	Ulna	Vertebral column					

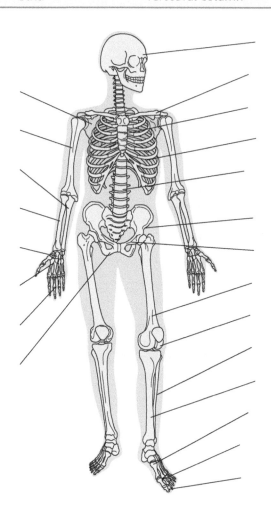

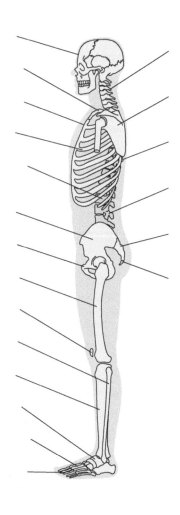

ACTIVITY 4: SORTING

Bones

Sort the following bones into the correct category.

Carpals   Femur   Fibula   Frontal   Humerus   Metacarpals   Patella
Ribs   Scapula   Sternum   Tarsals   Ulna   Vertebral column

Long	Short	Irregular	Flat	Sesamoid

Joints

Sort the following joints into the correct category.

Atlas and axis   Between vertebrae   Elbow   Hip   Knee   Radius and ulna   Shoulder   Thumb

Ball and socket	Hinge	Pivot	Saddle	Gliding

ACTIVITY 5: MATCH THE KEY WORDS

Match the following key terms to the correct description.

Axial   Compact   Femur   Ischium   Joint   Kyphosis
Ossification   Periosteum   Synovial   Tarsals

Key term	Description	Key term	Description
	Part of the skeleton consisting of the bones of the skull, vertebral column, sternum and ribs		Process of bone development
	Hard portion of the bone that makes up the main shaft of long bone		Freely movable joint
	Long bone of the thigh		Bones of the ankle
	Postural defect in which there is an abnormally increased outward curvature of the thoracic spine		Fibrous covering of a long bone
	Bone forming the inferior (lower) and posterior (back) part of pelvic girdle		Point at which two or more bones or cartilage meet

ACTIVITY 6: FILL IN THE BLANKS

Test your knowledge of the structure of bone by filling in the gaps with the terms shown below.

bone   calcium   cancellous   cartilage   centre   chondrocytes   compact
connective   embryo   ends   haversian   long   ossification   osteoblasts
osteoclasts   osteocytes   lighter   protein   red bone marrow   water

Bone is one of the hardest types of _____ tissue in the body and when fully developed is composed of _____, _____ and mineral salts.

Bone tissue is a type of living tissue that is made from special cells called _____.
There are two main types of bone tissue: _____ and cancellous.

Compact bone is the hard portion of the bone that makes up the main shaft of the _____ bones and the outer layer of other bones. It protects spongy bone and provides a firm framework for the bone and body.

The bone cells in this type of bone are called _____ and are located in concentric rings (called lamellae) around a central _____ canal, which consists of minute tubes that form a network in bone through which nerves, blood and lymphatic vessels pass.

In contrast, _____ bone is more porous and _____ in weight than compact bone. It has an open, sponge-like appearance and is found at the _____ of long bones or at the _____ of other bones. It does not have a haversian system but consists of a web-like arrangement of spaces that are filled with _____. Blood vessels run through every layer of cancellous bone, conveying nutrients and oxygen.

The process of bone development is called _____. The bones of a foetus are made of _____ rods that change into bone as the baby develops and grows. This process begins in the _____, near the end of the second month, and is not complete until about the 25th year of life. Ossification takes place in three stages. The cartilage-forming cells, called _____, enlarge and arrange themselves in rows similar to the bone they will eventually form. _____ salts are then laid down by special bone-building cells called osteoblasts.

A second set of cells called _____, known as cartilage-destroying cells, bring about an antagonistic action, enabling the absorption of any unwanted _____.

Chapter 5 The muscular system

INTRODUCTION

Having knowledge of the position and action of muscles allows therapists to be more accurate in their treatment applications to ensure effective results, as well as to recognise the varying degrees of muscle tone.

The essentials you need for exams and assessments

You need to know:

- **the structure and function of the three different types of muscle tissue** in the body: skeletal/voluntary, involuntary/smooth/non-striated and cardiac
- **the characteristics of muscle tissue**: contractibility, extensibility, elasticity and irritability; muscle tone and muscle tension, and how this may vary
- **how a muscle contracts**: including the role of actin and myosin, nerve stimulus, energy supply (glucose), aerobic, anaerobic and muscle fatigue
- **how a muscle provides movement**: including muscle attachments (tendons, ligaments, origins and insertions) and the co-ordination of muscle movement agonists (prime movers), antagonists, synergists and fixators (stabilisers)
- **position, origin and action of the muscles** of the head and face, posterior of trunk, anterior of trunk, arm and hand, leg and foot
- **functions of the muscular system**: movement, maintaining posture and heat production.

ACTIVITY 1: MULTIPLE-CHOICE QUESTIONS

1 The pectoralis major muscle is found in the:
   a shoulder
   b back
   c chest
   d abdomen.

2 Which of the following muscles is responsible for straightening the arm?
   a Brachialis
   b Triceps
   c Biceps
   d Deltoid

3 Skeletal muscle fibres are wrapped together in bundles known as:
   a fascia
   b epimysium
   c sarcolemma
   d fasciculi.

4 The type of muscle contraction that occurs when the muscle works without actual movement is:
   a isometric
   b isotonic
   c concentric
   d eccentric.

5 The term referring to muscles on the same side of a joint that work together to perform the same movement is:
   a agonist
   b protagonist
   c prime mover
   d synergist.

6 Muscles with less than the normal degree of tone are said to be:
   a spastic
   b rigid
   c flaccid
   d fibrotic.

7 The part that gives the skeletal muscle its striated or striped appearance when viewed under a microscope is:
   a actin and myosin
   b sarcomere and sarcolemma
   c fast twitch fibres
   d slow twitch fibres.

8 The muscle used when winking is the:
   a orbicularis oris
   b orbicularis oculi
   c zygomaticus
   d temporalis.

9 The name of the large triangular-shaped muscle in the upper back that raises the shoulder girdle is the:
   a splenius cervicus
   b trapezius
   c levator scapula
   d splenius capitis.

10 A chronic condition that produces musculoskeletal pain, lethargy and fatigue is:
   a fibrositis
   b myositis
   c fibromyalgia
   d muscular dystrophy.

Activity 2: Exam-style questions

1 Define the following terms in relation to muscle movement.   **3 marks**
  a Synergist   c Isometric contraction
  b Agonist

2 Describe the role of actin and myosin in muscle contraction.   **2 marks**

3 Name the fuel needed for muscle contraction.   **1 mark**

4 For each of the following muscles, give the position *and* one action.   **12 marks**
  a Frontalis   d Orbicularis oris
  b Corrugator   e Depressor anguli oris
  c Buccinator   f Risorius

5 For each of the following muscles, state their location in the body.   **6 marks**
  a Sternocleidomastoid   d Gastrocnemius
  b Trapezius   e Rectus abdominus
  c Deltoid   f Triceps

6 State one action of each of the following muscles.   **6 marks**
  a Brachioradialis   d Sartorius
  b Extensor carpi radialis   e Tibialis anterior
  c Quadriceps   f Flexor digitorum longus

ACTIVITY 3A: LABELLING THE MUSCLES OF THE HEAD AND NECK

Label the muscles of the head and neck using the words shown below.

Buccinator	Corrugator	Depressor anguli oris	Depressor labii inferioris
Frontalis	Levator anguli oris	Levator labii superioris	Masseter
Mentalis	Nasalis	Orbicularis oculi	Orbicularis oris
Platysma	Procerus	Risorius	Sternocleidomastoid
Temporalis	Zygomaticus major	Zygomaticus minor	

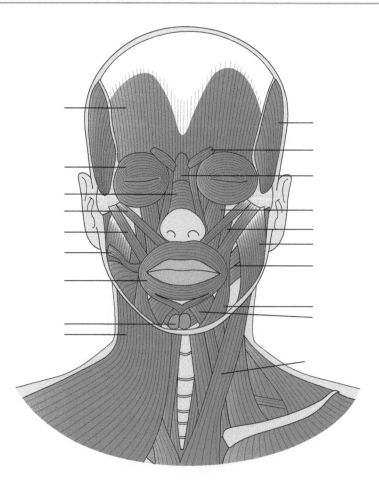

ACTIVITY 3B: LABELLING THE ANTERIOR MUSCLES OF THE BODY

Label the anterior muscles of the body using the words shown below.

Adductor
Deltoid
Internal oblique
Quadriceps
Vastus intermedius

Biceps
Extensor digitorum longus
Pectoralis major
Rectus abdominus
Tensor fasciae latae
Vastus lateralis

Coracobrachialis
External oblique
Pectoralis minor
Rectus femoris
Tibialis anterior
Vastus medialis

Flexors of forearm
Peroneus longus
Sartorius
Transversus abdominis

Surface muscles

Deep muscles

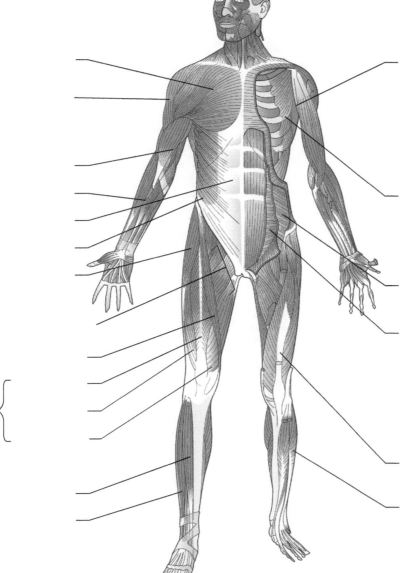

ACTIVITY 3C: LABELLING THE POSTERIOR MUSCLES OF THE BODY

Label the posterior muscles of the body using the words shown below.

Biceps femoris	Deltoid	Erector spinae	Extensors of forearm
Flexor digitorum longus	Flexor hallucis longus	Gastrocnemius	Gluteus maximus
Gluteus medius	Gluteus minimus	Hamstrings	Infraspinatus
Latissimus dorsi	Rhomboid major	Rhomboid minor	Rhomboids
Semimembranosus	Semitendinosus	Soleus	Supraspinatus
Tibialis posterior	Trapezius	Triceps	

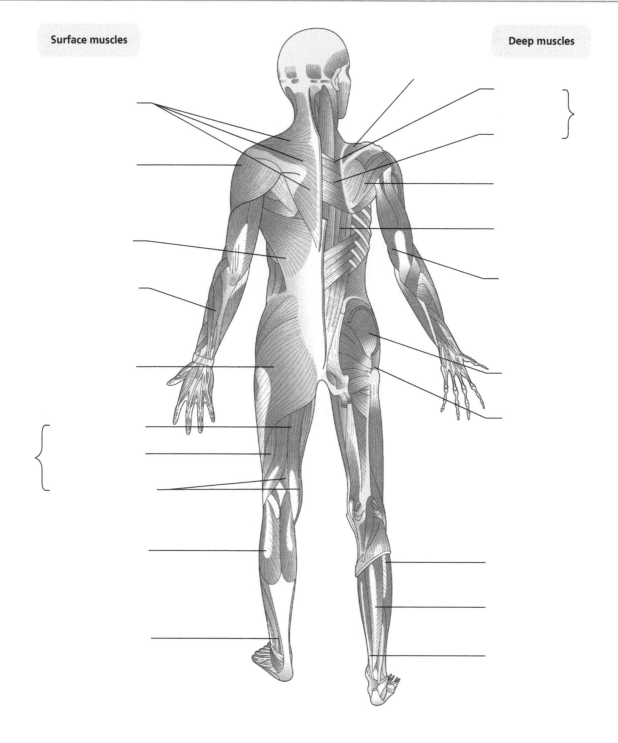

Surface muscles

Deep muscles

ACTIVITY 4: SORTING

Sort the following muscles into the correct category.

Abductor pollicis brevis
Corrugator
Extensor carpi ulnaris
Flexor hallucis longus
Latissimus dorsi
Nasalis
Peroneus tertius
Rhomboids
Sternocleidomastoid
Tibialis posterior

Biceps femoris
Depressor labii inferioris
External and internal obliques
Gastrocnemius
Masseter
Palmar aponeurosis
Quadratus lumborum
Sartorius
Supinator radii brevis
Transversus abdominis

Brachialis
Erector spinae
Flexor carpi radialis
Intercostals
Mentalis
Pectoralis minor and major
Rectus abdominis
Serratus anterior
Supraspinatus
Trapezius

Muscles of the head and face	Muscles of the arm and hand	Anterior muscles of the trunk	Posterior muscles of the trunk	Muscles of the leg and foot

ACTIVITY 5: MATCH THE KEY WORDS

Match the following key terms to the correct description.

| Cardiac | Fascia | Insertion | | Linea alba | Motor point |
| Muscle tone | Myofibrils | Neuromuscular junction | | Skeletal | Tendon |

Key term	Description	Key term	Description
	Type of muscle tissue attached to bone		Site where nerve fibres and muscle fibres meet
	Type of muscle tissue found in the walls of the heart		End part of nerve through which the stimulus to contract is given to the muscle fibre
	Contractile elements of a skeletal muscle fibre		Most movable part of the muscle during contraction
	Tough fibrous band that links muscle to bone		State of partial contraction of a muscle
	Fibrous connective tissue that envelops a muscle		Long tendon extending from the bottom of the sternum to the pubic symphysis

ACTIVITY 6: FILL IN THE BLANKS

Test your knowledge by filling in the gaps with the terms shown below.

| actin | calcium | elongate | motor | myosin |
| shorten | sliding | sodium | thicken | voluntary |

When a stimulus is applied to _____ muscle fibres via a _____ nerve, a mechanical action is initiated.

During contraction a _____ movement occurs within the contractile fibres of the muscle. The _____ protein filaments move inwards towards the _____ and the two filaments merge. This action causes the muscle fibres to _____ and _____, and then pull on their attachments (bones and joints) to effect the movement required. The attachment of myosin cross-bridges to actin requires _____.

The nerve impulses leading to contraction cause an increase in _____ ions within the muscle cell. During relaxation, the muscle fibres _____ and return to their original shape.

Chapter 6 The cardiovascular system

INTRODUCTION

The cardiovascular system is the body's transport system; it is comprised of blood, blood vessels and the heart. The heart acts like a pump that keeps the blood circulating around the body in a constant circuit. It is important for therapists to have a good working knowledge of the cardiovascular system in order to be able to understand how the treatments (such as massage) they provide help to improve circulation – the delivery of oxygen and nutrients to the tissues is improved, and the removal of waste products is hastened.

The essentials you need for exams and assessments

You need to know:

- **the composition of blood**: 55% plasma and 45% blood cells (erythrocytes, leucocytes and thrombocytes)
- **the blood groups**: A, B, AB and O
- **the functions of blood**: transport, defence, regulation of heat and clotting
- **the structure and function of blood vessels**:
  - arteries have thick muscular walls and carry blood away from the heart
  - veins have thinner walls, valves and carry blood towards the heart
  - capillaries are a single cell thick and unite arterioles and venules
- **the process of blood clotting (haemostasis)**: vasoconstriction, platelet plug formation/clot formation and the 12 clotting factors
- **blood pressure**: maximum pressure **systolic**, minimum pressure **diastolic**, and how it is measured (via a sphygmomanometer)
- **pulse rate**: pressure wave; can be felt in arteries; corresponds to the beating of the heart; how it is measured (may be felt and measured at a point where an artery lies near the surface, such as the radial artery in the wrist)
- **the structure and function of the heart**: made of three layers of cardiac tissue (pericardium, myocardium and endocardium); the heart is divided into right and left sides by a partition (septum); the upper chambers are the atria, the lower chambers are the ventricles; valves (atrioventricular bicuspid and tricuspid valves, semi-lunar valves, aortic and pulmonary valves); heart muscle is supplied by coronary arteries
- **the vessels involved in blood flow through the heart**: superior and inferior vena cavae, pulmonary arteries, pulmonary veins, aorta
- **the cardiac cycle**: sinoatrial node, atrioventricular node; cardiac centre in the medulla oblongata of the brain; role of the autonomic nervous system in controlling heart rate.
- **types of circulation**:
  - **pulmonary circulation** between the heart and the lungs
  - **systemic circulation** delivers oxygenated blood around the body through the various branches of the aorta
  - **portal circulation** collects blood from digestive organs and delivers it to the liver for processing via the hepatic portal vein
- **the main arteries of the face and head**: common carotid, external carotid, occipital, facial, maxillary, lingual, superficial temporal, thyroid
- **the main veins of the face and head**: external jugular, internal jugular, common facial, anterior facial, maxillary, superficial temporal
- **the main arteries of the body**: aorta, common carotid, subclavian, splenic, right and left iliac, renal artery, hepatic artery
- **the main veins of the body**: superior vena cava, inferior vena cava, splenic vein, right and left iliac vein, renal vein, hepatic vein, hepatic portal vein, subclavian
- **the main arteries of the arm**: subclavian, axillary, brachial, ulnar, radial, deep palmar arch, superficial palmar arch, digital arteries
- **the main veins of the arm**: subclavian, axillary, brachial, basilic, cephalic, ulnar, radial, palmar, digital veins
- **the main arteries of the leg**: external iliac, femoral, popliteal, anterior tibial, posterior tibial, plantar arch
- **the main veins of the leg**: long saphenous, short saphenous, dorsal venous arch, femoral, popliteal, anterior tibial, posterior tibial.

ACTIVITY 1: MULTIPLE-CHOICE QUESTIONS

1 The specialised blood cell involved in blood clotting is the:
   a lymphocyte
   b leucocyte
   c erythrocyte
   d platelet.

2 What is the function of the aorta?
   a to carry oxygenated blood around the body
   b to carry oxygenated blood to the heart
   c to carry deoxygenated blood to the lungs
   d to carry deoxygenated blood from the lungs

3 The heart muscle is supplied by the:
   a right and left carotid arteries
   b right and left coronary arteries
   c right and left iliac arteries
   d right and left occipital arteries.

4 The layer making up the bulk of the heart is the:
   a myocardium
   b pericardium
   c septum
   d endocardium.

5 What substance is removed from blood when it reaches the lungs?
   a carbon dioxide
   b oxygen
   c haemoglobin
   d hydrogen

6 The blood vessel supplying blood to the brain is the:
   a lingual artery
   b internal carotid artery
   c external carotid artery
   d maxillary artery.

7 The portal circulation collects blood from the digestive organs and delivers it to which organ for processing?
   a liver
   b spleen
   c pancreas
   d stomach

8 Blood pressure may be defined as the amount of pressure exerted by blood on an arterial wall due to the contraction of the:
   a aortic valve
   b vena cavae
   c left ventricle
   d left atrium.

9 The blood vessel that drains most of the blood from the face is the:
   a superficial temporal vein
   b internal jugular vein
   c external jugular vein
   d common facial vein.

10 An abnormal balloon-like swelling in the wall of an artery is an:
   a aneurysm
   b angina
   c arteriosclerosis
   d phlebitis.

11 What percentage of plasma is found in blood?
   a 40%
   b 50%
   c 55%
   b 80%

12 The sounds created by the beating heart are caused by:
   a blood moving from one chamber to another
   b compression from the respiring lungs
   c contraction of the ventricles
   d closing of the heart's valves.

Activity 2: Exam-style questions

1 a What is meant by clotting factor?   **1 mark**
  b Give two examples of clotting factors.   **2 marks**

2 Give two structural and one functional difference between each of the following blood vessels.
  a Artery   **3 marks**
  b Vein   **3 marks**

3 Explain the terms 'vasodilation' and 'vasoconstriction' in relation to blood capillaries.   **4 marks**

4 List the names of the layers of the heart from the outer to the inner layer.   **3 marks**

5 Define the following terms in relation to blood pressure.
  a Diastolic   **2 marks**
  b Systolic   **2 marks**

ACTIVITY 3: LABELLING THE STRUCTURE OF THE HEART

Label the structure of the heart using the words shown below.

Arch of aorta

Branches of ascending aorta

Left atrium

Left pulmonary veins

Right atrium

Right pulmonary veins

Superior vena cava

Bicuspid valve

Descending aorta

Left pulmonary artery

Left ventricle

Right pulmonary artery

Right ventricle

Tricuspid valve

Branch of pulmonary artery

Inferior vena cava

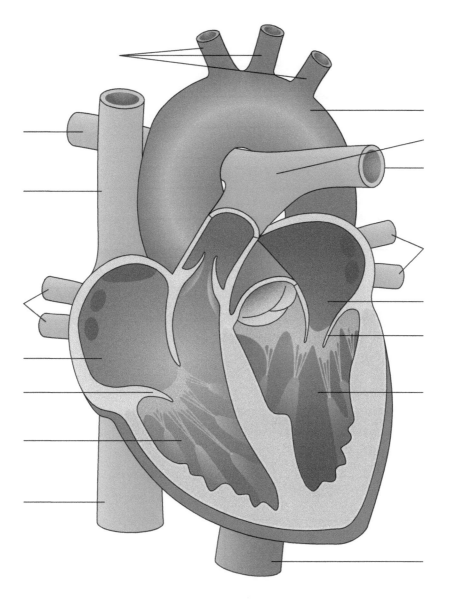

ACTIVITY 4: SORTING

Put the following stages of blood flow through the heart in the correct order.

The four pulmonary veins leave the lungs carrying oxygen-rich blood back to the left atrium.

When the left ventricle is full it contracts, forcing blood through the aortic valve into the aorta and to all parts of the body (except the lungs).

When the right atrium is full, it empties through the tricuspid valve into the right ventricle.

Deoxygenated blood from the body enters the superior and inferior vena cava and flows into the right atrium.

When the right ventricle is full, it contracts and pushes blood through the pulmonary valve into the pulmonary artery.

The pulmonary artery divides into the right and left branch and takes blood to both lungs, where the blood becomes oxygenated.

Oxygen-rich blood leaves the left atrium and passes through the left ventricle via the bicuspid (or mitral) valve.

1	
2	
3	
4	
5	
6	
7	

ACTIVITY 5: MATCH THE KEY WORDS

Match the following key terms to the correct description.

Artery	Capillary	Fibrin	Haemostasis
Inferior vena cava	Leucocyte	Lumen	Septum
Superior vena cava	Thrombocyte/platelet	Vein	Ventricle

Key term	Description	Key term	Description
	Smallest type of blood vessel that unites arterioles and venules		Blood vessel carrying blood towards the heart
	Specialised blood cell involved in clotting		Largest of all the blood cells
	Blood vessel carrying blood away from the heart		Large vein draining deoxygenated blood from the upper parts of the body above the diaphragm (head, neck, thorax and arms)
	The process by which bleeding is stopped		An opening inside a blood vessel through which blood flows
	Insoluble protein that forms a fibrous mesh during blood clotting		Large vein receiving deoxygenated blood from the lower parts of the body below the diaphragm
	Partition separating the two chambers of the heart		One of the two lower chambers of the heart

ACTIVITY 6: FILL IN THE BLANKS

Test your knowledge of the composition and function of blood by filling in the gaps with the terms shown below.

bacteria	blood	carbon dioxide	clot	clotting	disease	glucose
heat	hormones	infection	lactic	leucocytes	liver	lymphocytes
muscles	oxygen	pH	phagocytes	thrombocytes	transport	urea

Blood is the primary _____ medium for a variety of substances that travel throughout the body. _____ is carried from the lungs to the cells of the body in red blood cells, and _____ is carried from the body's cells to the lungs. Other substances carried in the blood include nutrients such as _____, amino acids, vitamins and minerals, and cellular wastes such as water, _____ acid and _____ to be excreted.

_____, which are internal secretions that help to control important body processes, are transported by the blood to target organs.

White blood cells are collectively called _____. They play a major role in combating _____ and fighting _____.
_____ have the ability to engulf and ingest micro-organisms that invade the body and cause disease. Specialised white blood cells called _____ produce antibodies to protect the body against infection.

Blood also helps to regulate _____ in the body by absorbing large quantities produced by the _____ and the _____. This is then transported around the body to help to maintain a constant internal temperature. Blood also helps to regulate the body's _____ balance.
_____ is an effective mechanism in controlling blood loss from blood vessels when they have become damaged, as in a cut. Specialised blood cells called _____, or platelets, form a _____ around the damaged area to prevent the body from losing too much _____ and to prevent the entry of _____.

Chapter 7 The lymphatic system and immunity

INTRODUCTION

The lymphatic system is a unidirectional (one-way) drainage system for the tissues. It provides a circulatory pathway for tissue fluid to be transported as lymph from the tissue spaces of the body into the venous system, where it becomes part of the blood circulation.

It is important for therapists to have a working knowledge of the lymphatic system in order to understand the effects of lymphatic drainage on the tissues. Any treatments that relax the soft tissue, such as massage, can help to accelerate lymph drainage as it will encourage the muscles to relax and the lymphatic vessels to open.

The essentials you need for exams and assessments

You need to know:
- **the structure and function of the parts of the lymphatic system** including:
  - lymph (interstitial fluid)
  - lymphatic capillaries, which drain excess fluid from tissue spaces
  - lymphatic vessels, which carry lymph towards the heart and drain into at least one node
  - lymphatic nodes, which contain lymphocytes and filter lymph
  - lymphatic ducts (thoracic and right lymphatic), which collect lymph before returning it to the blood circulation via the subclavian veins
- **the position of the main lymph nodes of the face** including buccal, mandibular, mastoid, occipital, submental, submandibular, parotid (anterior auricular) and mastoid (posterior auricular)
- **the position of the main lymph nodes of the body** including axillary, supratrochlear, inguinal, popliteal, superficial and deep cervical, cisterna chyli and other lymphatic organs (thymus, spleen and tonsils)
- **the movement of lymph/lymphatic drainage**: unidirectional (one-way) drainage relies on skeletal/muscular contractions and changes in internal pressure during respiration – it has no pump
- **functions of the lymphatic system**: defence against invading micro-organisms, maintaining the correct balance of body fluids, absorption of fats (lacteals).

ACTIVITY 1: MULTIPLE-CHOICE QUESTIONS

1 The lymphatic system has a close relationship to which other system?
   a nervous
   b respiratory
   c circulatory
   d renal

2 The two main lymphatic ducts are the:
   a right and left subclavian
   b thoracic and left subclavian
   c thoracic and right lymphatic
   d right and left lymphatic.

3 The largest of the lymphatic organs is the:
   a tonsils
   b spleen
   c liver
   d thymus.

4 Collected lymph is drained into the venous system via the:
   a subclavian arteries
   b superior vena cava
   c brachiocephalic veins
   d subclavian veins.

5 Which of the following is **not** a lymphatic organ?
   a thymus
   b tonsils
   c pancreas
   d spleen

6 What is the function of the lymphatic capillaries?
   a to carry lymph back to the bloodstream
   b to carry excess fluid towards the tissues of the body
   c to carry excess fluid away from the tissue spaces
   d to carry lymph towards the heart

7 The submental lymphatic nodes are located:
   a behind the ear
   b under the chin
   c on the underside of the jaw
   d in the cheek.

8 The function of the cisterna chyli is to:
   a drain lymph laden with digested fat from the small intestine
   b filter lymph that has accumulated in the abdomen
   c stop infection accumulating in the abdomen
   d drain lymph laden with infection from the small intestine.

9 The vessels that enter a lymphatic node are described as:
  a capsular
  b nodular
  c efferent
  d afferent.

10 A malignant disease of the lymphatic tissues, usually characterised by painless enlargement of one or more groups of lymph nodes in the neck is:
  a acquired immune deficiency syndrome (AIDS)
  b Hodgkin's disease
  c Lupus erythematosus
  d none of the above.

Activity 2: Exam-style questions

1 List the four structures that make up the lymphatic system.   **4 marks**
2 State two functions of the lymphatic system.   **2 marks**
3 Where are the following lymph nodes located?   **2 marks**
  a Submental   **b** Mastoid
4 List two features of lymphatic vessels.   **2 marks**
5 State the area(s) of the body that the thoracic duct collects lymph from, stating which blood vessel it drains into before being returned to the bloodstream.   **2 marks**

ACTIVITY 3A: LABELLING THE MAIN LYMPH NODES OF THE FACE

Label the position of the main lymph nodes of face using the words shown below.

Buccal nodes	Deep cervical nodes	Mastoid nodes	Occipital nodes
Parotid nodes	Submandibular nodes	Submental nodes	Superficial cervical nodes

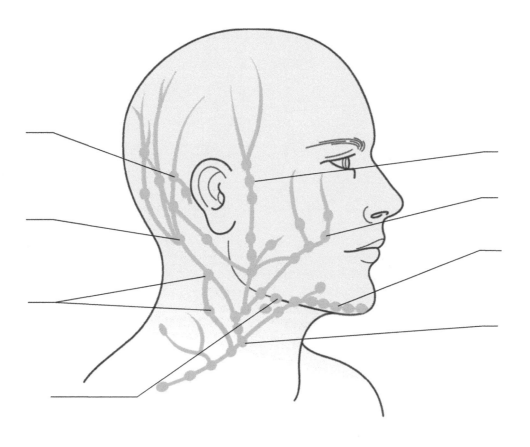

ACTIVITY 3B: LABELLING THE MAIN LYMPH NODES OF THE BODY

Label the position of the main lymph nodes of body using the words shown below.

Abdominal nodes	Axillary nodes	Cervical nodes	Cisterna chyli
Cubital/supratrochlear nodes	Inguinal nodes	Pelvic nodes	Popliteal nodes
Right lymphatic duct	Thoracic duct	Thoracic nodes	

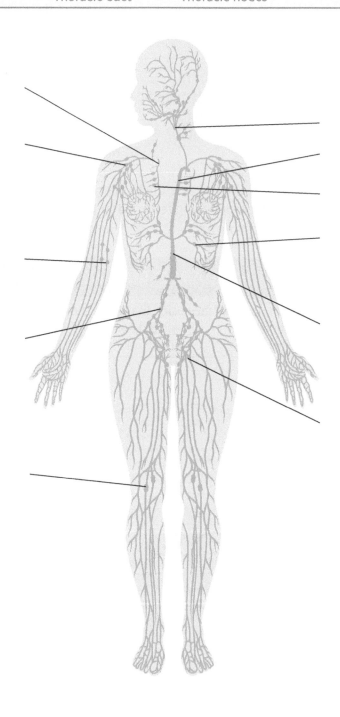

ACTIVITY 4: SORTING

Test your knowledge by putting the stages of the circulatory pathway of lymph in the correct order below.

Larger lymphatic vessels lead to lymph nodes.
Excess fluid flows through a network of lymphatic capillaries.
Tissue fluid enters lymph vessels where it becomes lymph.
Collected lymph is drained into the venous system via the subclavian veins.
Lymph passes through at least one lymphatic node where it is filtered.
Plasma escapes blood capillaries and bathes tissue cells.
Filtered lymph is collected into lymphatic ducts.

1	
2	
3	
4	
5	
6	
7	

ACTIVITY 5: MATCH THE KEY WORDS

Match the following key terms to the correct description.

Cisterna chyli   Lymph (lymphatic fluid)   Lymphatic capillary   Lymphatic duct   Lymphatic node
Lymphatic vessel   Lymphocyte   Mastoid   Oedema   Spleen

Key term	Description	Key term	Description
	Transparent, colourless, watery liquid derived from tissue fluid		Tube that transports lymph through its circulatory pathway
	Specialised type of white blood cell		Minute, blind-end tube that drains excess fluid and waste from the tissue spaces of the body
	Largest of the lymphatic organs		Excess fluid within the tissue spaces that causes the tissues to become waterlogged
	Collecting vessel of the lymphatic system		Lymphatic nodes behind the ear
	Oval or bean-shaped structure that filters lymph of micro-organisms		Drains lymph laden with digested fats from the intestine

ACTIVITY 6: FILL IN THE BLANKS

Test your knowledge of the lymphatic system by filling in the gaps with the terms shown below.

blood plasma   carbon dioxide   liquid   lymph   lymphocytes   oxygen
plasma   proteins   respiration   skeletal   subclavian   tissue fluid
urea   veins   vessels   water

Lymph is a transparent, colourless, watery _____ and is contained within lymphatic
_____. It resembles _____ _____in composition, except that it has a lower concentration
of plasma _____. Lymph contains only one type of cell – these are called _____.
As blood is distributed to the tissues, some of the _____ escapes from the capillaries and flows
around the tissue cells, delivering nutrients such as _____ and _____, and picking up
cellular waste such as _____ and _____ _____. Once the plasma is outside the capillary and
is bathing the tissue cells, it becomes _____ _____. Some of the tissue fluid passes back into the
capillary walls to return to the bloodstream via the _____, and some is collected by lymph vessels,
where it becomes _____.

As the lymphatic system lacks a pump, lymphatic vessels have to make use of _____ muscles
contracting, and changes in internal pressure during _____, to assist the movement of lymph.
Lymph is then taken through its circulatory pathway and is ultimately returned to the bloodstream via the
_____ veins.

Chapter 8 The respiratory system

INTRODUCTION

Respiration is the process by which the living cells of the body receive a constant supply of oxygen and remove carbon dioxide and other gases. The respiratory system also plays a role in detecting smell, producing speech and regulating pH.

It is important for therapists to have a good knowledge of the respiratory system to understand how breathing may be affected during a treatment.

In general, treatments of a relaxing nature will tend to deepen respiration and improve lung capacity by relaxing any tightness in the respiratory muscles.

The essentials you need for exams and assessments

You need to know:

- **the structure and function of the organs of the respiratory system** including the nose, naso-pharynx, pharynx, larynx, trachea, bronchi, bronchioles, lungs and alveoli
- **external respiration**: gas exchange that takes place in the lungs between the blood and air in the alveoli that came from the external environment
- **internal respiration**: gas exchange between the blood and the tissues throughout the body
- **mechanism of respiration**: diaphragm and intercostal muscles
- **nervous control of respiration**: chemoreceptors and the respiratory centre in the medulla oblongata of brain
- **modified respiratory movements** including crying, coughing, hiccups, laughing, sighing, sneezing, talking and yawning.

ACTIVITY 1: MULTIPLE-CHOICE QUESTIONS

1 Which of the following is the chief muscle of respiration?
   a external intercostal
   b internal intercostal
   c pectoralis major
   d diaphragm

2 During gas exchange, oxygen and carbon dioxide diffusion occurs in the:
   a red blood cells   c alveoli
   b venules   d body tissues.

3 When oxygen passes through the alveoli into the bloodstream it binds with haemoglobin to form:
   a red blood cells   c carbon dioxide
   b oxyhaemoglobin   d nitrogen.

4 Involuntary breathing results from stimulation of the respiratory centre in the:
   a cerebellum
   b thalamus
   c medulla and pons
   d hypothalamus.

5 What prevents dirt from entering the lungs?
   a cartilage   c cilia
   b mucus   d air

6 At the back of the nasopharynx there is lymphoid tissue called the:
   a tonsils   c epiglottis
   b adenoids   d Adam's apple.

7 What is the function of the pleural cavity in the lungs?
   a to enable a more efficient exchange of gases
   b to transport air from the bronchioles into the lungs
   c to allow the diaphragm to contract more effectively
   d to prevent friction between the visceral and parietal layers

8 The tonsils are found at the back of the:
   a larynx
   b pharynx
   c trachea
   d adenoids.

9 When is the breathing rate likely to decrease?
   a during exercise
   b at times of extreme stress
   c during sleep
   d when emotions are running high

10 A chronic obstructive pulmonary disease in which the alveoli of the lungs become enlarged and damaged is:
   a asthma
   b emphysema
   c pleurisy
   d pneumonia.

Activity 2: Exam-style questions

1	State two functions of the respiratory system.	**2 marks**
2	State the muscles involved in the mechanism of respiration.	**3 marks**
3	Define what is meant by the term 'interchange of gases'.	**2 marks**
4	What prevents dirt from entering the lungs?	**1 mark**
5 a	What is the name given to the tiny air-filled sacs in the lungs?	**1 mark**
b	What do these air-filled sacs provide for the exchange of gases?	**1 mark**

ACTIVITY 3: LABELLING THE RESPIRATORY SYSTEM

Label the structures of the respiratory system using the words shown below.

Bronchioles	Bronchus	Epiglottis	Intercostal muscle	Larynx
Left lung	Nasopharynx	Oesophagus	Pharynx	Pleural cavity
Pleural membrane	Rib	Right lung	Trachea	

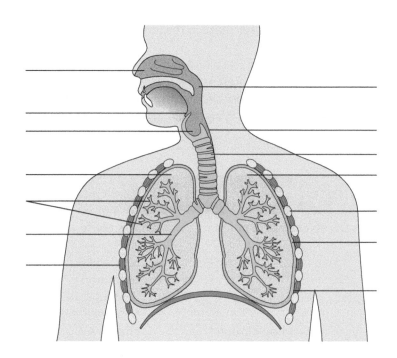

ACTIVITY 4: SORTING

Place the following respiratory structures into the order that air would pass through them during a normal inspiration.

Bronchi   Bronchioles   Larynx   Lungs   Nasopharynx   Pharynx   Trachea

1	
2	
3	
4	
5	
6	
7	

ACTIVITY 5: MATCH THE KEY WORDS

Match the following key terms to the correct description.

Alveoli   Bronchi   Diaphragm   Larynx   Lungs   Mucus   Nose   Pharynx
Sinuses   Trachea

Key term	Description	Key term	Description
	Box-like cavity containing the vocal cords		Cone-shaped spongy organs situated on either side of the heart
	Tube made of cartilage that transports air from the larynx into the bronchi		Tiny air sacs in the lungs
	Two short tubes that lead to and carry air into each lung		Dome-shaped muscular partition separating the thoracic cavity from the abdominal cavity
	Organ that moistens, warms and filters incoming air, and senses smell		Sticky fluid that prevents dust and bacteria from entering the lungs
	Large muscular tube that acts as a passageway for air, food and drink		Air-filled spaces located within the maxillary, frontal, ethmoid and sphenoid bones of the skull

ACTIVITY 6: FILL IN THE BLANKS

Test your knowledge of the mechanism of respiration by filling in the gaps with the terms shown below.

air   contract   depth   diaphragm   external intercostal   lungs   medulla oblongata
passive   pons   recoil   relaxation   respiratory   thoracic cavity   thorax

The mechanism of respiration is the means by which _____ is drawn in and out of the
_____.

It is an active process where the muscles of respiration _____ to increase the volume of the
_____ _____. During inspiration the _____ contracts and flattens, increasing the volume of
the thoracic cavity. It is responsible for 75% of air movement into the lungs. The _____ _____ muscles
are also involved in respiration: on contraction they increase the _____ of the thoracic cavity by
pulling the ribs upwards and outwards. They are responsible for bringing approximately 25% of the volume of
air into the lungs.

During normal respiration, the process of expiration is _____ and is brought about by the
_____ of the diaphragm and the external intercostal muscles, along with the elastic
_____ of the lungs. This increases the internal pressure inside the _____, so that air
is pushed out of the lungs.

Breathing is controlled by a group of neurones in the parts of the brain called the _____ _____ and the
_____ (known as the _____ centre).

Chapter 9 The nervous system

INTRODUCTION

The nervous system is the body's control centre. It is responsible for receiving and interpreting information from inside and outside the body.

It is important for therapists to have a comprehensive knowledge of the nervous system in order to be able to understand the effects of treatments. Some treatments may have the ability to stimulate nerves while others have the ability to relax.

Knowledge of the nervous system can also help therapists to understand the effects of stress on the body.

The essentials you need for exams and assessments

You need to know:

- **the structure and function of neurones**: a neurone is a specialised type of nerve cell; there are three types of neurones – sensory nerve (afferent), motor nerve (efferent), mixed nerve
  - **structure**: nerve cell body, axon and dendrites, as well as myelin sheath, neurilemma, nodes of Ranvier, synapse, synaptic cleft, axon ending/terminals and the neurotransmitters acetylcholine and noradrenaline
  - **function**: designed to receive stimuli and conduct impulses
- **characteristics of nervous tissue**: irritability and conductivity
- **the structure and function of the central nervous system (CNS)**:
  - **structure**: CNS is covered by meninges – pia mater, arachnoid mater and dura mater; cerebrospinal fluid protects the CNS; the CNS has two types of tissue – grey matter and white matter
  - **brain**: cerebrum, thalamus, hypothalamus, cerebellum, mid-brain, medulla oblongata, pons
  - **spinal cord**: extension of the brain stem, function is to relay impulses to and from the brain
  - **reflex arc/action**: rapid and automatic response to a stimulus without any conscious thought of the brain
- **the structure of the peripheral system, face and neck**: contains all the nerves outside of the CNS – 12 pairs of cranial nerves and 31 pairs of spinal nerves
  - **network of nerves**: plexus
- **the autonomic nervous system**: sympathetic and parasympathetic
- **the five senses**: smell, sight, hearing, taste and touch.

ACTIVITY 1: MULTIPLE-CHOICE QUESTIONS

1 What part of the neurone transmits impulses away from the cell body?
  a myelin sheath
  b synapse
  c axon
  d dendrites

2 The region of the brain concerned with the co-ordination of skeletal muscle is the:
  a cerebellum
  b pons
  c mid-brain
  d cerebrum.

3 The spinal cord is an extension of which part of the brain?
  a pons
  b medulla oblongata
  c mid-brain
  d brain stem

4 Which of the following is **not** one of the cranial nerves?
  a trigeminal
  b facial
  c cervical
  d optic

5 The effects of the parasympathetic nervous system are:
  a resting heart rate
  b increased gastrointestinal activity
  c pupil constriction
  d all of the above.

6 The fine, delicate membrane that surrounds the axon is called the:
  a neuroglia
  b neurilemma
  c nodes of Ranvier
  d synaptic knob.

7 The part of the neurone that covers the axon, insulating and accelerating the conduction of nerve impulses along the length of the axon, is called the:
  a nodes of Ranvier
  b dendrite
  c myelin sheath
  d motor end plate.

8 The part of the brain that contains vital control centres for the heart, lungs and intestines is the:
  a hypothalamus
  b mid-brain
  c cerebellum
  d medulla oblongata.

9 The junction where nerve impulses are transmitted from one neurone to another is a:
  a neurotransmitter
  b synapse
  c dendrite
  d axon.

10 A disease of the central nervous system in which the myelin (fatty) sheath covering the nerve fibres is destroyed and various functions become impaired is:
  a Parkinson's disease
  b epilepsy
  c multiple sclerosis
  d neuralgia.

Activity 2: Exam-style questions

1 State two functions of the nervous system.   2 marks
2 State the two main parts of the nervous system.   2 marks
3 a Define what is meant by a neurone.   2 marks
  b State the three basic parts of a neurone's structure.   3 marks
4 State one difference between a sensory nerve and a motor nerve.   2 marks
5 Name the largest part of the brain.   1 mark

ACTIVITY 3: LABELLING THE PRINCIPAL PARTS OF THE BRAIN

Label the principal parts of the brain using the words shown below.

| Brain stem | Cerebellum | Cerebrum | Hypothalamus | Medulla oblongata |
| Mid-brain | Pineal gland | Pons | Spinal cord | Thalamus |

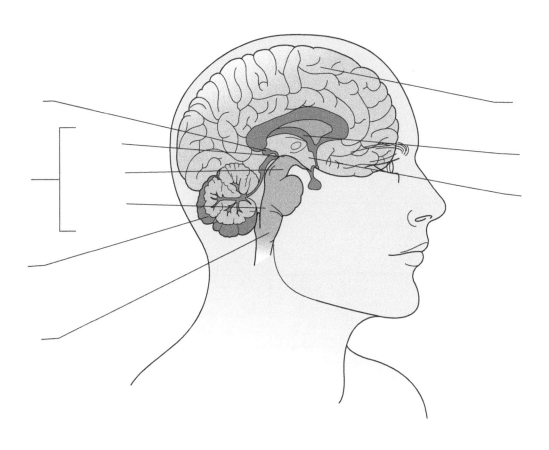

ACTIVITY 4: SORTING

Sort the following effects of the autonomic nervous system into the correct category.

Increases breathing rate	Constricts blood vessels	Constricts pupils
Contracts bladder	Decreases secretion of saliva	Dilates blood vessels
Dilates pupils	Increases conversion of	Increases conversion of
Increases secretion of saliva	glucose to glycogen	glycogen to glucose by liver
Increases the secretion of sweat	Increases heart rate	Increases peristalsis
Slows down breathing rate	Reduces peristalsis	Relaxes bladder
	Slows down heart rate	Stimulates release of adrenalin

Effects of sympathetic nervous system stimulation	Effects of parasympathetic nervous system stimulation

ACTIVITY 5: MATCH THE KEY WORDS

Match the following key terms to the correct description.

Acetylcholine   Afferent   Autonomic nervous system   Central nervous system   Cerebellum
Dendrites   Neurone   Noradrenaline   Peripheral nervous system   Reflex action

Key term	Description	Key term	Description
	Main control system consisting of brain and spinal cord		Rapid and automatic response to a stimulus without any conscious thought of the brain
	System of nerves that link the central nervous system to the rest of the body		Highly branched extensions of the nerve cell
	Part of the nervous system that controls the automatic body activities of smooth and cardiac muscle and the activities of glands		Neurotransmitter that causes muscles to contract, activates pain responses and regulates endocrine/sleep functions
	Functional unit of the nervous system		Part of brain concerned with muscle tone, the co-ordination of skeletal muscles and balance
	Type of nerves that receive stimuli from sensory organs and receptors, and transmit the impulse to the spinal cord and brain		Neurotransmitter designed to mobilise the brain and body for action

ACTIVITY 6: FILL IN THE BLANKS

Test your knowledge of the autonomic nervous system by filling in the gaps with the terms shown below.

cardiac   digestion   dilates   emergency   energy   gastrointestinal
genito-urinary   glands   heart   parasympathetic respiration   rest
saliva   sleep   slows   smooth   sweat   sympathetic

The autonomic nervous system controls the automatic body activities of _____ and _____ muscle and the activities of _____. It is divided into the sympathetic and parasympathetic systems.

The _____ system prepares the body for expending _____ and dealing with _____ situations. It increases the _____ and _____ rate, _____ skeletal blood vessels, stimulates the _____ and adrenal glands, dilates the pupils, decreases the secretion of _____, and decreases _____ activity.

The _____ nervous system works to conserve energy and create the conditions needed for _____ and _____. It _____ down the body processes, except _____, and the functions of the _____ system.

Chapter 10 The endocrine system

INTRODUCTION

The endocrine system is comprised of a series of internal secretions called hormones that help to regulate body processes by providing a constant internal environment.

It is important for therapists to have a comprehensive knowledge of the endocrine system in order to understand the action of hormones and their significance in the healthy functioning of the body. Over- or under-secretion of certain hormones may result in disorders and disease in the body.

The essentials you need for exams and assessments

You need to know:

- **the structure of endocrine glands**: ductless glands; the hormones they secrete pass directly into the bloodstream to influence the activity of another organ or gland; they have a feedback mechanism co-ordinated by the hypothalamus and the pituitary gland
- **the position of the endocrine glands**: hypothalamus (brain), pituitary (brain), pineal (brain), thyroid (neck, either side of trachea), parathyroids (posterior of thyroid), islets of Langerhans (pancreas), adrenals (on top of kidneys), ovaries (lower abdomen), testes (groin)
- **the function of hormone(s) secreted by the endocrine glands**:
  - **pituitary gland**: growth hormone, thyroid-stimulating hormone (TSH), adrenocorticotrophic hormone (ACTH), gonadotrophic hormones (follicle-stimulating hormone, luteinising hormone), prolactin, melanin-stimulating hormone, oxytocin, antidiuretic hormone (vasopressin)
  - **adrenal gland (cortex)**: mineral corticoids, glucocorticoids and sex corticoids
  - **adrenal gland (medulla)**: adrenaline, noradrenaline
  - **thyroid gland**: thyroxine (T4), triiodothyronine (T3), calcitonin
  - **parathyroid glands**: parathormone
  - **islets of Langerhans**: insulin (beta cells), glucagon (alpha cells), somatostatin (gamma cells)
  - **pineal gland**: melatonin
  - **thymus gland**: thymosin
- **disorders of the endocrine glands and their symptoms**:
  - **pituitary disorders**: gigantism, acromegaly, Simmonds' disease, Lorain-Levi syndrome, Fröhlich's syndrome, diabetes insipidus
  - **thyroid disorders**: Graves' disease (thyrotoxicosis), congenital iodine deficiency syndrome, myxoedema
  - **parathyroid disorders**: Cushing's syndrome, Addison's disease, adrenal hyperplasia
  - **islets of Langerhans**: diabetes mellitus
  - **ovaries**: polycystic ovarian syndrome (Stein-Leventhal syndrome)
- **the role and effects that hormones have on natural glandular changes**: puberty, pregnancy, menopause, menstrual cycle and stress.

ACTIVITY 1: MULTIPLE-CHOICE QUESTIONS

1   The effects of adrenaline are:
    a   increased heart rate
    b   increased metabolic rate
    c   increased oxygen intake
    d   all of the above.

2   The glucocorticoids are secreted by the:
    a   adrenal medulla   c   pancreas
    b   adrenal cortex   d   pineal gland.

3   In puberty the ovaries are stimulated by:
    a   gonadotrophic hormones from the anterior pituitary

    b   growth hormone from the anterior pituitary
    c   prolactin from the anterior pituitary
    d   adrenaline from the adrenal medulla.

4   Hypersecretion of testosterone in women can lead to the condition:
    a   gynaecomastia
    b   polycystic ovary syndrome
    c   amenorrhea
    d   menopause.

5 An increased metabolic rate, weight loss, sweating and restlessness are all symptoms of:
  a thyrotoxicosis   c myxoedema
  b Cushing's   d diabetes
    syndrome   mellitus.

6 Endocrine glands in the body have a feedback mechanism that is co-ordinated by which gland?
  a pituitary   c adrenals
  b thyroid   d pineal

7 The thymus serves a vital role in the development of:
  a neutrophils
  b basophils
  c T lymphocytes
  d B lymphocytes.

8 The pea-sized mass of nerve tissue attached by a stalk in the central part of the brain is the:
  a hypothalamus   c pineal gland
  b pituitary gland   d corpus callosum.

9 The role of the hormone calcitonin is to control:
  a the level of thyroxine in the blood
  b the level of calcium in the blood
  c metabolism
  d heart and digestive function.

10 Hypersecretion of the hormone thyroxine leads to a condition known as:
  a myxoedema
  b Simmonds' disease
  c Cushing's syndrome
  d thyrotoxicosis.

Activity 2: Exam-style questions

1 Name the hormone that stimulates the muscles of the uterus during childbirth.   **1 mark**

2 State the location of the following endocrine glands: **3 marks**
  a Hypothalamus   **b** Thyroid   **c** Adrenals

3 What is the name of the hormone that increases water reabsorption in the renal tubules of the kidneys?   **1 mark**

4 State three functions of the thyroid gland.   **3 marks**

5 State two effects of the hormone adrenaline on the body.   **2 marks**

ACTIVITY 3: LABELLING THE ENDOCRINE GLANDS

Label the endocrine glands using the words shown below.

Adrenal   Ovary   Pancreas (islets of Langerhans)   Pineal
Pituitary   Testis   Thyroid

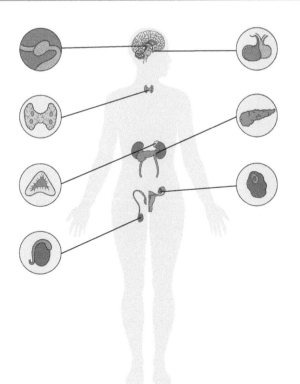

ACTIVITY 4: SORTING

Sort the following hormones into their correct category, according to the endocrine gland that secretes them.

Adrenaline
Calcitonin
Glucocorticoids
Insulin
Mineral corticoids
Oxytocin
Prolactin
Somatostatin
Thyroid-stimulating hormone (TSH)
Triiodothyronine (T3)

Adrenocorticotrophic hormone (ACTH)
Follicle-stimulating hormone (FSH)
Growth hormone
Luteinising hormone (LH)
Noradrenaline
Parathormone
Relaxin
Testosterone

Antidiuretic hormone (ADH)
Glucagon
Inhibin
Melanocyte-stimulating hormone (MSH)
Oestrogen
Progesterone
Sex corticoids
Thymosin
Thyroxine (T4)

Anterior lobe of pituitary gland	
Posterior lobe of pituitary gland	
Thyroid gland	
Parathyroid glands	
Thymus	
Adrenal cortex	
Adrenal medulla	
Pancreas	
Ovaries	
Testes	

ACTIVITY 5: MATCH THE KEY WORDS

Match the following key terms to the correct description.

Adrenal glands   Alpha cells   Beta cells   Islets of Langerhans   Melatonin
Noradrenaline   Parathyroid glands Pituitary gland   Thymosin   Thyroid gland

Key term	Description	Key term	Description
	Lobed structure attached by a stalk to the hypothalamus of the brain		Endocrine tissue located within the pancreas
	Gland in neck responsible for controlling metabolism		Hormone produced by the adrenal medulla when the body is at rest
	Triangular-shaped glands that lie on top of each kidney		Cells in the pancreas responsible for producing the hormone insulin
	Four small glands situated on the posterior of the thyroid		Cells in the pancreas responsible for producing the hormone glucagon
	Hormone of the thymus gland		Hormone produced by the pineal gland

ACTIVITY 6: FILL IN THE BLANKS

Test your knowledge of the endocrine system by filling in the gaps with the terms shown below.

Anterior   Corpus luteum   Endometrium   Fallopian tube
Follicle   Follicle-stimulating hormone Gonadotrophic   Luteinising hormone
Menstrual   Menstrual cycle   Menstruation   Ovaries
Ovulation   Puberty   Ovum   Pituitary
Pregnancy   Progesterone   Proliferative   Reproductive
Secretory   Testes   Uterus

_____ is the time at which the onset of sexual maturity occurs and the _____ organs become functional.

Changes are brought about by an increase in sex hormone activity, due to stimulation of the _____ and _____ by the pituitary _____ hormones.

Starting at puberty, the female reproductive system undergoes a regular sequence of monthly events known as the _____ _____.

There are three stages of the cycle, which lasts approximately 28 days. The first stage (days 7–14) is the _____ stage, days 14–28 is the _____ phase and the third phase (days 1–7) is the _____ phase.

At the beginning of the cycle an _____ develops within an ovarian _____ in the ovaries in response to a hormone released by the anterior lobe of the _____ gland called the _____ _____, which stimulates the follicles of the ovaries to produce the hormone oestrogen.

When mature, the ovum bursts from the follicle and travels along the _____ _____ to the _____. This occurs about 14 days after the start of the cycle and is known as _____.

A temporary endocrine gland, the _____ _____, develops in the ruptured follicle in response to stimulation from the _____ _____ secreted by the _____ lobe of the pituitary gland.

This temporary gland secretes the hormone _____ which, together with oestrogen, causes the lining of the uterus to become thicker and richly supplied with blood in preparation for _____.

After ovulation, the ovum can only be fertilised during the next 8–24 hours. If fertilisation does occur, the fertilised ovum becomes attached to the endometrium and the corpus luteum continues to secrete progesterone. Pregnancy then begins. If the ovum is not fertilised, the cycle continues – the corpus luteum shrinks and the _____ (lining of uterus) is shed. This is called _____. The cycle then begins again.

Chapter 11 The reproductive system

INTRODUCTION

The reproductive systems are the only systems that are very different for men and women, both in terms of structure and function.

It is important for therapists to have a comprehensive knowledge of the reproductive system in order to be able to understand the effects of the natural glandular changes in the body.

The essentials you need for exams and assessments

You need to know:

- **the structure and function of the female reproductive system:**
  - **structure**: vulva (mons pubis, labia majora and minora, clitoris, greater vestibular glands, vulval vestibul), vagina, uterus, cervix, fallopian tubes, ovaries, ova, urinary meatus
  - **function**: fertilisation/gestation/birth
- **the structure and function of the male reproductive system:**
  - **structure**: testes, vas deferens, epididymis, prostate gland, scrotum, penis
  - **function**: sperm production and urination
- **cycles and the hormones involved**: puberty, menstrual cycle, ovulation, pregnancy, menopause and hormone replacement therapy
- **the structure and function of the breast:**
  - **structure**: made of glandular and adipose/areolar connective tissues; supported by suspensory ligaments (Cooper's), pectoralis major and serratus anterior muscles; milk ducts open into nipple (surrounded by pigmented areola); nipple contain glands of Montgomery and sensory nerve endings; lymphatic drainage is mainly into axillary nodes; blood supply to breast is the internal thoracic artery, a branch of the subclavian artery and the axillary arteries; the veins of the breast drain into the axillary and internal thoracic veins
  - **function**: milk (colostrum) production, involving the hormones prolactin and oxytocin, for lactation following childbirth.

ACTIVITY 1: MULTIPLE-CHOICE QUESTIONS

1 The female genitalia is known collectively as the:
   a vulva
   b vagina
   c vestibule
   d vestibular gland.

2 During menstruation, which part of the uterus is shed?
   a myometrium
   b endometrium
   c perimetrium
   d none of the above

3 Where does fertilisation of the ovum take place?
   a ovaries
   b uterus
   c fallopian tubes
   d vagina

4 The absence or stopping of menstrual periods is known as:
   a endometriosis
   b dysmenorrhea
   c amenorrhea
   d premenstrual syndrome.

5 The ejaculatory ducts are:
   a a pair of ducts that open into the urethra at the base of the penis
   b two short tubes that join the seminal vesicle to the urethra
   c composed of erectile tissue
   d tubes leading from the epididymis to the urethra.

6 The primary function of the testes is:
   a to store seminal fluid
   b to produce and maintain sperm cells
   c to develop the male secondary sexual characteristics
   d to nourish immature sperm cells.

7 The first secretion from the mammary glands after giving birth is called:
   a colostrum
   b crura
   c choroid
   d none of the above.

8 The hormone responsible for the development of the secondary sexual characteristics in women is:
   a oxytocin
   b oestrogen
   c progesterone
   d prolactin.

9 A condition that presents with inflammation of the inner lining of the uterus, abnormal menstrual bleeding and lower abdominal pain is:
   a ectopic pregnancy
   b endometriosis
   c fibroids
   d ovarian cyst.

10 A foetus's sex is distinguishable at:
   a eight weeks
   b the end of the first month
   c the end of the first trimester
   d a month before birth.

11 An abnormal growth of fibrous and muscular tissue, one or more of which may develop in the muscular wall of the uterus, is:
   a endometriosis
   b fibroids
   c polycystic ovary syndrome
   d dysmenorrhea.

Activity 2: Exam-style questions

1 Name the part of the male reproductive system where immature sperm cells are stored.   **1 mark**
2 What is the term given to a fertilised ovum?   **1 mark**
3 Name the part of the female reproductive system where fertilisation of the ovum takes place.   **1 mark**
4 State the dual function of the urethra in men.   **2 marks**
5 a What is the name of the tubes leading from the epididymis to the urethra?   **1 mark**
   b What is their function?   **1 mark**

ACTIVITY 3A: LABELLING THE FEMALE REPRODUCTIVE ORGANS

Label the female reproductive organs using the words shown below.
Cervix   Fallopian tube   Ovaries   Ovum   Uterus   Vagina

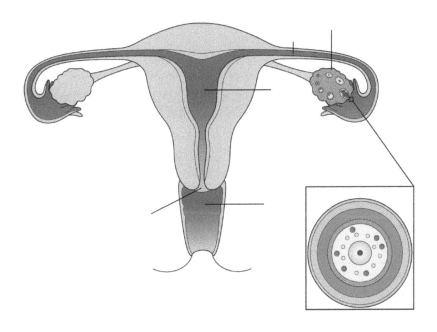

ACTIVITY 3B: LABELLING THE MALE REPRODUCTIVE ORGANS

Label the male reproductive organs using the words shown below.

Epididymis   Penis   Prostate gland   Seminal vesicle   Testis   Ureter   Urethra   Vas deferens

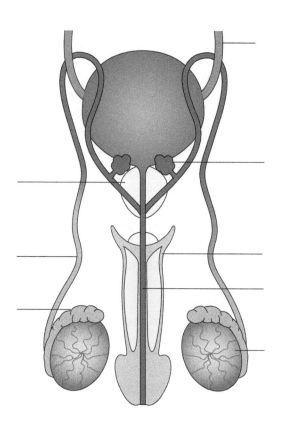

ACTIVITY 4: MATCH THE KEY WORDS

Match the following key terms to the correct description.

Endometrium   Labia majora and minora   Ovaries   Ovum   Progesterone   Testes
Urinary meatus   Uterus   Vagina   Vas deferens

Key term	Description	Key term	Description
	Egg cell		Lip-like folds at the entrance of the vagina
	Muscular organ in which embryo grows		Lining of the uterus that is shed each month during menstruation
	Female sex glands		Stimulates the thickening of the uterine lining in anticipation of implantation of a fertilised egg
	Muscular and elastic tube that provides a passageway for menstruation and childbirth		Tubes through which sperm is released
	Male reproductive glands		Vertical slit located at the tip of the glans penis

ACTIVITY 5: FILL IN THE BLANKS

Test your knowledge by filling in the gaps with the terms shown below.

Atrophy   Cervix   Elasticity   Fatty   Gonadotrophic   Libido   Lubrication
Menopause   Mucus   Oestrogen   Ovaries   Progesterone   Prostate   Seminal
Sperm   Testes   Testosterone   Vagina   Viscous   Vulva

Numerous physical changes occur in women with increasing age as the levels of the hormones _____ and _____ decline.

Ovulation usually stops between one and two years before the _____. As the _____ reach the end of their productive cycle, they become unresponsive to _____ stimulation. With age, the ovaries _____ and become thicker and smaller.

The _____ also atrophies with age and the tissue shrinks. Atrophy causes the _____ to shorten and the _____ lining to become thin, dry and less elastic.

After the menopause the uterus shrinks rapidly to half its premenstrual weight. The _____ atrophies and no longer produces mucus for _____.
In the breasts, the glandular, supporting and _____ tissues atrophy and, as the Cooper's ligaments lose their _____, large breasts become pendulous.

Physiological changes in older men include reduced _____ production, which in turn may cause decreased _____. A reduced testosterone level also causes the _____ to atrophy and soften, and decreases _____ production by around 48–69% between the ages of 60 and 80.

Normally, the _____ gland enlarges with age and its secretions diminish. _____ fluid also decreases in volume and becomes less _____.

Chapter 12 The digestive system

INTRODUCTION

In the digestive system food is broken down and made soluble before it can be absorbed by the body for nutrition. Food is taken in through the mouth, broken into smaller particles and absorbed into the bloodstream, where it is utilised by the body.

It is important for therapists to have a good knowledge of the process of digestion in order to understand how the body utilises nutrients for efficient and healthy body functioning.

The essentials you need for exams and assessments

You need to know:

- **the digestive processes** including ingestion, digestion (mechanical and chemical), absorption, assimilation and elimination/defaecation
- **the structure and function of the digestive system from the ingestion of food to the elimination of waste**: mouth, tongue (papillae), teeth, pharynx, epiglottis, salivary glands, oesophagus, cardiac sphincter, stomach, pyloric sphincter, small intestine (duodenum, jejunum and ileum, ileocaecal valve), large intestine (caecum, ascending colon, transverse colon, descending colon, sigmoid colon, rectum) and anal canal
- **the mechanical and chemical breakdown of food**: digestive juices/enzymes involved in the breakdown of carbohydrates, protein and fats, such as:
  - **saliva**, containing the enzyme salivary amylase, or ptyalin, commences the digestion of starch, or carbohydrates, in the mouth
  - **pepsin** (the precursor of pepsinogen) is the main gastric enzyme that starts the breakdown of proteins into polypeptides in the stomach (along with hydrochloric acid, mucus and water)
  - **bile** breaks down fat droplets via emulsification
  - **cholecystokinin (CCK)**: hormone responsible for stimulating the digestion of fats and protein
  - **secretin**: hormone released into the bloodstream by the duodenum (especially in response to acidity) to stimulate secretion by the liver and pancreas
  - **pancreatic proteases**: produced by the pancreas but released into the small intestine
    - **enteropeptidase** (also called enterokinase) is an enzyme that converts trypsinogen into its active form, trypsin
    - **trypsin and chymotrypsin** help to digest proteins; protein digestion is completed by peptidases, which split short-chain polypeptides into amino acids
  - **pancreatic amylase** helps to digest sugars (carbohydrates); carbohydrate digestion is completed by maltase (which splits maltose into glucose, sucrase (which splits sucrose into glucose and fructose) and lactase (which splits lactose into glucose and galactose)
  - **pancreatic lipase** helps to digest fats (monoglycerides, diglycerides and triglycerides) into fatty acids and glycerol
- **the structure and function of the accessory digestive organs**: the liver, gall bladder and pancreas
- **the dietary components/main food groups required for good health**: carbohydrates, proteins, fats, water, minerals, vitamins and roughage.

ACTIVITY 1: MULTIPLE-CHOICE QUESTIONS

1 Where in the digestive tract does peristalsis occur?
   a only in the mouth
   b only in the small intestine
   c only in the stomach
   d in all sections of the alimentary canal

2 The enzyme that commences the digestion of starch, or carbohydrates, in the mouth is:
   a pepsin
   b pancreatic amylase
   c salivary amylase
   d enterokinase.

3 The function of hydrochloric acid secreted into the stomach is to:
   a curdle milk proteins
   b break down large fat particles
   c commence the breakdown of proteins
   d neutralise the germs in food.

4 Vitamins and minerals are absorbed into the bloodstream via the:
   a liver cells
   b villi in the small intestine
   c lacteals in the small intestine
   d hepatic portal vein.

5 Which of the following is responsible for producing bile?
   a liver
   b gall bladder
   c pancreas
   d duodenum

6 In which part of the large intestine are faeces stored before defaecation?
   a caecum
   b rectum
   c descending colon
   d appendix

7 The colon primarily absorbs which substance?
   a carbohydrates   c fats
   b proteins   d water

8 The hormone released by cells in the duodenum in response to chyme that has a high fat or protein content is called:
   a enterokinase
   b secretin
   c cholecystokinin
   d rennin.

9 Pancreatic proteases that help to digest proteins are:
   a monoglycerides and diglycerides
   b trypsin and chymotrypsin
   c polysaccharides and disaccharides
   d polypeptides and tripeptides.

10 A condition that presents with abnormal dilatation of veins in the rectum is:
   a haemorrhoids   c hernia
   b heartburn   d jaundice.

11 The main part of the large intestine is the:
   a duodenum
   b caecum
   c colon
   d ileum.

12 The three sections of the small intestine, from beginning to end, are:
   a ascending, transverse and descending
   b jejunum, ileum and duodenum
   c duodenum, jejunum and ileum
   d duodenum, ileum and jejunum.

13 Which enzyme starts the digestion of proteins in the stomach?
   a trypsin
   b maltase
   c pepsin
   d gastrin

14 Glucose, the end product of carbohydrate digestion, is used to:
   a provide energy for cells to function
   b produce new tissues
   c repair damaged cell parts
   d build new cell membranes.

Activity 2: Exam-style questions

1 State three constituents of gastric juice, which is produced in the stomach.   **3 marks**
2 List two enzymes produced and secreted by the pancreas.   **2 marks**
3 Name the parts of the large intestine.   **4 marks**
4 What is the name given to the special features of the small intestine into which nutrients pass to enter the bloodstream?   **1 mark**
5 In what part of the digestive system is the chemical digestion of food completed?   **1 mark**

ACTIVITY 3: LABELLING THE DIGESTIVE ORGANS

Label the digestive organs using the words shown below.

Anal canal	Anal sphincter	Appendix	Caecum	Colon	Duodenum
Gall bladder	Ileocaecal valve	Ileum	Jejunum	Liver	Mouth
Oesophagus	Pancreas	Pyloric sphincter	Rectum	Small intestine	Stomach
Salivary glands					

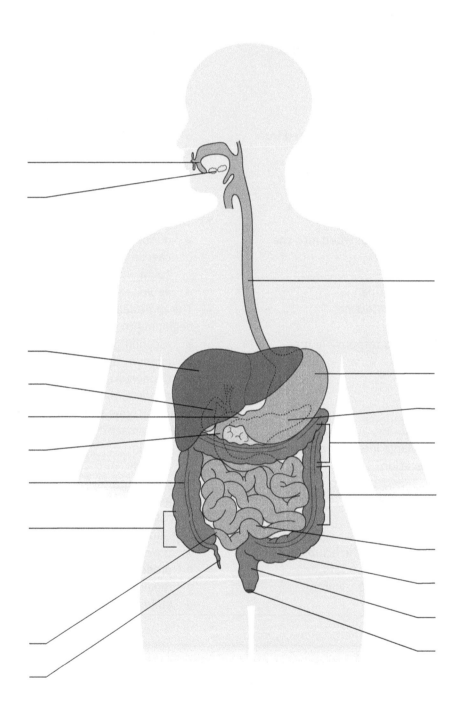

ACTIVITY 4A: SORTING

Place the following organs in the order they are involved in the digestive process.

Anus Ascending colon Caecum Descending colon Duodenum Ileum Jejunum
Mouth Oesophagus Pharynx Rectum Sigmoid colon Stomach Transverse colon

1	8
2	9
3	10
4	11
5	12
6	13
7	14

ACTIVITY 4B: SORTING

Sort the following digestive enzymes into the correct category according to where they are produced.

Enteropeptidases (enterokinase) Lactase Maltase Pancreatic amylase
Pancreatic lipase Pepsin Salivary amylase Sucrase
Trypsin and chymotrypsin

Mouth	Stomach	Pancreas	Small intestine

ACTIVITY 5: MATCH THE KEY WORDS

Match the following key terms to the correct description.

Absorption   Anus   Colon   Defaecation   Digestion   Enzyme   Ingestion
Liver   Oesophagus   Peristalsis

Key term	Description	Key term	Description
	Chemical catalyst that activates and speeds up a chemical reaction		Main part of the large intestine
	The process of breaking down food		Opening at the lower end of the alimentary canal
	The act of taking food into the alimentary canal through the mouth		The expulsion of semi-solid waste called faeces through the anal canal
	Co-ordinated rhythmical contractions of the circular and oblique muscles in the wall of the alimentary tract		Largest organ in the body
	Long, narrow tube linking the pharynx to the stomach		Movement of soluble materials out through the walls of the small intestine

ACTIVITY 6: FILL IN THE BLANKS

Test your knowledge of digestion by filling in the gaps with the terms shown below.

amino acids   anus   bile   chyme   faeces   gastric   glucose   intestinal
lacteals   liver   mouth   oesophagus   pancreatic   pepsin   peristalsis
protein   rectum   saliva   small intestine   starch   villi   vitamins

Digestion commences in the _____, where food is chewed by the teeth and mixed thoroughly with _____, which contains an enzyme called amylase that starts to digest _____.

Food then passes down the _____ to the stomach. It is conveyed by a process of rhythmic muscular contractions called _____.

In the stomach the food is churned up and mixed with _____ juice containing the enzyme _____, which starts to digest _____.

The food stays in the stomach for approximately five hours, until it has been churned down into a liquid state called _____.

Food is then passed into the _____ _____ where more enzymes continue the chemical breakdown of food.

The food is also mixed with _____, which is manufactured in the liver to help emulsify fat, _____ juice from the pancreas to continue the digestion of protein and carbohydrates, and _____ juice, which completes the final breakdown of nutrients (including simple sugars) to _____ and protein to _____ _____.

The absorption of digested food occurs by diffusion through the _____ of the small intestine, which are small, finger-like projections that are well supplied with blood capillaries.

_____ and minerals are absorbed in the blood capillaries while products of fat digestion are absorbed into the lymphatic system by lymph vessels called _____.

The capillaries join to form the hepatic portal vein, which transports the digested food to the _____ to be regulated before being utilised by the body's tissues.

Undigested food passes into the colon where a large amount of water is absorbed. The solid, undigested matter, known as _____, passes into the _____ where it is stored before being passed out of the body through the _____.

Chapter 13 The renal system

INTRODUCTION

It is essential for therapists to have a working knowledge of the renal system in order to understand how fluid balance is controlled in the body and the role of the kidneys in detoxification.

The essentials you need for exams and assessments

You need to know:

- **the structures of the urinary system**: kidneys, ureters, urinary bladder, urethra
- **the functions of the urinary system**: maintain homeostasis by controlling the composition, volume and pressure of blood, and regulating the composition and volume of body fluids
- **the structure of the kidneys**: the outer cortex and inner medulla contain tiny blood filtration units called nephrons that process filtered liquid that eventually becomes urine; inside the kidney the renal artery splits into a network of capillaries called the glomerulus; high pressure forces fluid out through the walls of the glomerulus into the surrounding Bowman's capsule (filtration); the filtered liquid continues through a series of twisted tubes called the proximal convoluted tubules (where substances are selectively reabsorbed), to the loop of Henle and then the distal convoluted tubules (where some substances are secreted) and on to the renal pelvis of the kidney; from here urine passes through the ureter and into the bladder; it is excreted through the urethra via micturition
- **the formation of urine in a nephron**: filtration, selective reabsorption and secretion/collection
- **the composition of urine**: 96% water, 2% urea, 2% other substances such as salts (chloride, sodium, potassium), creatinine, other dissolved ions, inorganic and organic compounds (proteins, hormones, metabolites)
- **the factors that affect urine production**: weather (hot/cold), exercise/activity, inactivity, stress, water consumption and medication
- **the hormones affecting urine production**: ADH/vasopressin, aldosterone, calcitonin.

ACTIVITY 1: MULTIPLE-CHOICE QUESTIONS

1   The outside part of a kidney is called the:
   a   hilus
   b   medulla
   c   cortex
   d   pelvis.

2   Which part of the renal system stores urine?
   a   ureter
   b   bladder
   c   urethra
   d   kidney

3   Which of the following factors affect fluid balance?
   a   diet
   b   body temperature
   c   blood pressure
   d   all of the above

4   Inside the kidney, the renal artery splits into a network of capillaries called the:
   a   afferent arteriole
   b   efferent arteriole
   c   hilus
   d   glomerulus.

5   The part of the kidney where the fluid is filtered from blood is the:
   a   renal pyramid
   b   medulla
   c   cortex
   d   hilus.

6   You are feeling dehydrated as you have not drunk enough fluids. How do your kidneys respond?
   a   release ADH
   b   release aldosterone
   c   inhibit the release of ADH
   d   increase urine output

7   Which of the following surrounds the glomerulus?
   a   distal convoluted tubule
   b   Bowman's capsule
   c   proximal convoluted tubule
   d   renal pelvis

8   Which of the following is **not** a function of the kidneys?
   a   to produce urine
   b   to filter blood
   c   to store urine
   d   to reabsorb substances needed by the body

9 The area where the renal blood vessels leave and enter the kidney is called the:
   a hilus   c calcyx
   b pelvis   d pyramid.

10 An inflammation of the urinary bladder, usually caused by an infection of the bladder lining, is:
   a incontinence   c cystitis
   b nephritis   d kidney stones.

Activity 2: Exam-style questions

1 What is the anatomical position of the kidneys?   **2 marks**
2 Describe the structure of a kidney.   **2 marks**
3 Describe two differences between the afferent and efferent arterioles carrying blood to and from the kidney.   **2 marks**
4 What is the process of expelling urine from the bladder called?   **1 mark**
5 Which part of the kidney filters waste?   **1 mark**

ACTIVITY 3: LABELLING THE RENAL SYSTEM

Label the urinary system using the words shown below.

| Bladder | Cortex | Left kidney | Medulla | Renal artery | Renal pelvis | Renal vein |
| Right kidney | Ureter | Urethra | | | | |

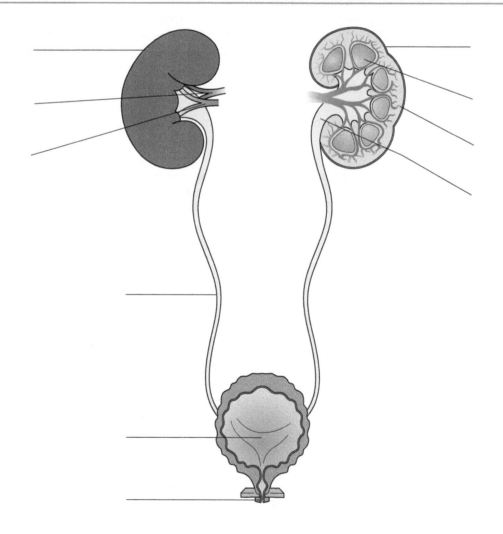

ACTIVITY 4: MATCH THE KEY WORDS

Match the following key terms to the correct description.

Aldosterone   Bladder   Calyces (calcyx)   Cortex   Kidney   Medulla
Nephrons   Renal pyramids   Ureter   Urethra

Key term	Description	Key term	Description
	Main functional organ of urinary system		Tiny blood filtration units inside the kidney
	Long tube that transports urine from the kidney to the bladder		Inner part of kidney's structure
	Pear-shaped sac that stores urine		Conical-shaped sections inside the medulla of the kidney
	Tube through which urine is discharged from the bladder to the exterior		Cup-shaped cavities in which urine collects
	Outer part of kidney's structure		Hormone that regulates the reabsorption of sodium and water in the kidneys

ACTIVITY 5: FILL IN THE BLANKS

Test your knowledge of urine production by filling in the gaps with the terms shown below.

afferent   amino acids   bloodstream   Bowman's capsule
convoluted tubules   distal convoluted tubules   efferent   filters
filtration   glomerulus   glucose   loop of Henle
medulla   micturition   proximal convoluted tubules   renal artery
renal pelvis   ureter   urethra   urine

The blood that needs to be processed enters the _____ of the kidney from the _____
_____. Inside the kidney the renal artery splits into a network of capillaries called the _____,
which _____ the waste. Almost encasing the glomerulus lies a sac called the _____ _____.
The blood pressure in the glomerulus is maintained at a high level, assisted by the fact that the
_____ arteriole feeding into the glomerulus has a larger diameter than the _____
arteriole leaving it. This pressure forces fluid out through the walls of the glomerulus, together with some
of the substances of small molecular size able to pass through the capillary walls into the Bowman's capsule.
This process constitutes simple _____.
The filtered liquid continues through a series of twisted tubes called the _____ _____, which are
surrounded by capillaries. The tubules of the nephron that lead away from the Bowman's capsule are known as
the _____ _____, which straighten out to form a long loop called the _____ _____. There
are then another series of twists called the _____ _____.
The composition of the filtered liquid alters as it flows through the convoluted tubules. Some substances
contained within the waste, such as _____, _____, mineral salts and vitamins, are
reabsorbed back into the _____ as the body cannot afford to lose them.
Excess water, salts and the waste product urea are all filtered and processed through the kidneys. The treated
blood leaves the kidney via the renal vein.
The wastes remaining in the distal convoluted tubule, now known as _____, then flow on via a
collecting tubule to the _____ _____ of the kidney. From here it passes into the _____
and on to the bladder and _____ to be excreted through a process known as _____.

Answers and extra activities can be found on: www.hoddereducation.co.uk/Anatomy-and-Physiology-Extras

Orders: Hachette UK Distribution, Hely Hutchinson Centre, Milton Road, Didcot, Oxfordshire, OX11 7HH. Telephone: +44 (0)1235 827827. Email education@hachette.co.uk Lines are open from 9 a.m. to 5 p.m., Monday to Friday. You can also order through our website: www.hoddereducation.co.uk

ISBN: 978 1 5104 3613 8

© Helen McGuinness 2018

First published in 2018 by
Hodder Education,
An Hachette UK Company
Carmelite House
50 Victoria Embankment
London EC4Y 0DZ

www.hoddereducation.co.uk

Impression number   10 9 8 7 6 5

Year   2023

Cover photo © Sebastian Kaulitzki – stock.adobe.com

Illustrations by Barking Dog Art.

Typeset in India.

Printed at Ashford Colour Press Ltd

A catalogue record for this title is available from the British Library.

HODDER EDUCATION

t: 01235 827827

e: education@hachette.co.uk

w: hoddereducation.co.uk

ISBN 978-1-5104-3613-8

MIX
Paper | Supporting
responsible forestry
FSC™ C104740